12/06

ALBUQUERQUE ACADEMY
1955

This book was donated
to honor
the birthday of

Santiago Leyba

by

Orlando + Kelley Leyba

2005

DATE

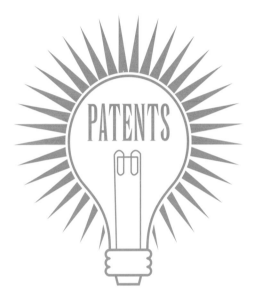

PATENTS

Ingenious Inventions
How They Work and How They Came to Be

BEN IKENSON

BLACK DOG
& LEVENTHAL
PUBLISHERS
NEW YORK

ISBN 1-57912-367-8
Library of Congress Cataloging-in-Publication Data
Ikenson, Ben.
Patents: ingenious inventions—how they work and how they came to be
Ben Ikenson.-- 1st ed.
    p. cm.
Includes bibliographical references and index.
ISBN 1-57912-367-8
1. Technology--Popular works. 2. Inventions. I. Title.

T47.I44 2004
608--dc22

2004000155

Book design: Scott Citron
Manufactured in China

Published by
Black Dog & Leventhal Publishers, Inc.
151 West 19th Street
New York, New York 10011

Distributed by
Workman Publishing Company
708 Broadway
New York, New York 10003

g f e d c b a

ACKNOWLEDGMENTS

Many people are to be acknowledged in the making of this book. First,
thanks to my editor, Laura Ross, whose creativity is as impressive as her
motivational skill. Reeve Chace provided superb preliminary text support.
Copyeditor/Fact Checker/Photo Researcher/Life Saver Sylvia Helm, along
with Dan Armstrong and Stuart Armstrong, offered remarkable assistance
throughout. The prodigious design talents of Scott Citron brought the book
to life, and my publisher, J. P. Leventhal, along with many others at Black
Dog, helped this labor bear fruit.

I would also like to thank a number of people who supported my efforts in
one way or another: Melanie Ruiz, Billy Kiester, Mark Olague, Tamara Ward,
Matt Huggler, Patrick Durham, Betsy Lordan, Doug Hobbs, Mary Leonard,
Alan Rothschild, and the professionals at the U.S. Patent and Trade Office.

608
IKE

2004

TABLE OF CONTENTS

INTRODUCTION

"Hell, there are no rules here—we're trying to accomplish something."

—Thomas A. Edison

ROM THE FIRST WHEEL WAY BACK WHEN, our inventions have been
what move us forward. Our evolution still depends largely on the tools
we create, great ideas made manifest, improved upon, occasionally perfected. Everywhere, we live in a world of ideas materializing.

As of this writing, a yellow bulldozer/backhoe hulks outside of my apartment window, large trucks beep in reverse; metal blades shear through pave-
ment; bulldozers scoop giant concrete chunks from the street as easily as a
boy scoops up a handful of marbles; a construction manager shouts into a
cell phone. They are replacing old waterlines that were surely an innovation
in their day. When it's all said and done, the people living on this block will
shower and wash, safe from the threat of lead poisoning from the old piping. In the meantime, I throw a CD into the stereo to drown out the ruckus
and sit down at the laptop computer to tap out flurries of e-mail over the
Internet. I do not have far to go to find inspiration for a book about human
ingenuity.

As civilizations emerged, so did systems of economy that rewarded innovation, ultimately attempting to protect that abstraction so crucial to any
alleged meritocracy, "intellectual property." Today, these are called patent and
trade offices, and many countries have established sophisticated systems of
laws which they help to uphold. A patent protects a person's idea so that he
might rightly profit from it, thereby encouraging innovation as a means to
prosperity.

In this book, many ideas, large and small, are explored and celebrated.
Inventions tend to reflect our needs and desires, sometimes even our fears;
they are inspired by the drive to improve life, make it more manageable, more
efficient, and even more fun. If necessity is the mother of invention, then
madcap ingenuity must be its errant father. How does one explain the success
of the Slinky or the perennial attraction of the Chia Pet without careening
happily towards the absurd?

Four close friends, Thomas Edison, John Burroughs, Henry Ford, and Harvey Firestone, inspect an antique mill wheel.

As a broad survey, this book celebrates all branches of the patent family tree. Be it Bubble Wrap, barbed wire, or the artificial heart, a patent reveals our values, our idiosyncrasies, and the spirit of invention that is such a fundamental part of human nature. This illustrated collection of patents offers insight into some of the defining principles of each invention represented, the inventor's original intention (sometimes wildly different from its ultimate use), and the peculiar visionary genius these singular patents were issued to protect. Many of the items were patented in an inventor's home country in addition to the United States. For the sake of consistency, this book relies mostly on the patent language and illustrations obtained from the U.S. Patent Office.

Since Thomas Jefferson handed out the first patent in 1790, the United States Patent and Trademark Office has granted more than six and a half million patents in the effort to foster scientific advancement and economic prosperity. These pages represent only a tiny yet significant fraction of these.

I hope this book does more than reflect upon the particular genius of some well-known objects and ideas; I hope it stirs within the reader that innate desire to invent. Like evolution, one invention often gives rise to another—the replacement of the water pipes beneath my street is a good example of this. That I can once again hear the ruckus outside because my old CD player is skipping is yet another. Invention is a constant work-in-progress to which this book pays respectful tribute.

—*Ben Ikenson*

PATENTS: A HISTORY

"Congress shall have the power to promote the progress of science and useful arts by securing for limited times to authors and inventors the exclusive right to their respective writings and discoveries."—Article, Section 8, U.S. Constitution

The Beginning

Although many of the United States's original thirteen colonies upheld some form of patent law, the original concept of the patent, as we now know it, was not a uniquely American idea. In 1449, King Henry VI of England awarded a patent to a "John" of "Utynam" for his distinctive way of manufacturing stained glass. It soon became clear that offering inventors some protection of their ideas benefited not only the individuals who came up with new things, but helped contribute to the economic growth of whole countries. So, when the settlers from England began arriving in America, they brought the concept of patent law with them.

So strong was the belief of the founding fathers in the usefulness of patent, trademark, and copyright law that tenets of the principle were written into the Constitution: Article 1, Section 8, Clause 8 of the U.S. Constitution, the "Intellectual Property Clause," is the basis upon which our patent laws are built. In fact, the contemporary patent office is founded on three acts passed during our country's early years: the Patent Act of 1790, the Patent Act of 1793, and the Patent Act of 1836.

Thomas Jefferson was among the group that led the charge to establish the first patent laws in the United States in 1790. The Patent Act of 1790 stipulated that all applications must be accompanied by a model of the invention, since Jefferson felt that patents should only be issued for tangible, physical things, not ideas. The first patent act also explicitly prohibited foreign patents from winning protection on U.S. soil. Samuel Hopkins of Pittsford, Vermont, was the recipient of the very first U.S. patent awarded by the new office for his improvement in the making of "Pot Ash."

A few years later, a second act was passed, partly in response to the large number of complaints from inventor about the inefficiencies of the first act. In 1793, a compromise of two competing bills drafted by Thomas Jefferson and Alexander Hamilton was drawn up, resulting in the Patent Act of 1793. This act established a formal review board consisting of three positions, the Secretary of State, Attorney General, and Secretary of War. However, there still was little organization within the patent office.

Reinventing the Patent Office

A rise in patent applications after the Civil War and during Reconstruction firmly established the need for a more efficient patent office. The Patent Act of 1836 established a Patent Office under the Department of State, and Henry Ellsworth, who helped draft the Act, was appointed the first Commissioner of

Patents. In addition to streamlining the application procedure, the Patent Act of 1836 also called for copies of all new patents to be distributed to libraries throughout the country, providing the general public with knowledge about the latest inventions. It is this act that the current Patent Office is based on.

In 1975, the Patent Office changed its name to the Patent and Trademark Office. Today it is now one of fourteen bureaus of the Department of Commerce and is located in Arlington, Virginia, where more than five thousand people are employed by the office for the express purpose of examining patent and trademark applications.

Model Trouble

On December 15, 1836, a fire in the Patent Office destroyed all records and most of the patent models. Congress appropriated $100,000 for the restoration of 3,000 of the most important models. Forty years later, a second fire destroyed another 76,000 models. In 1880, the model requirement was deemed impractical, and in 1893, the remaining models were placed in storage.

Eventually, the models were sold at auction or became lost to obscurity. Some 2,500 of them ended up at the Smithsonian Institution, while thousands more landed in the estate of philanthropist Sir Henry Wellcome who, it is believed, intended to establish a museum for them. Following his death, many of these were in turn auctioned off. In 1979, Cliff Petersen, a designer and inventor within the aerospace industry, obtained some eight hundred crates—some of which had not been opened since they were packed up in 1926—for $500,000. Petersen donated 30,000 models and one million dollars to the United States Patent Model Foundation, keeping approximately 5,000 models in his personal collection—yet many historic models were still landing in antique shops and flea markets.

A Home for Patent Models

In 1998, Alan Rothschild created the Rothschild Petersen Patent Model Museum in Cazenovia, New York. A significant segment of his collection was purchased from Cliff Petersen, and Rothschild also purchased all eighty-two models comprising the Patent Model Museum in Fort Smith, Arkansas. Thanks to the generosity of the Rothschild Petersen Patent Model Museum, several photographs of these models are represented in this book. ❧

FRANZ VESTER, OF NEWARK, NEW JERSEY.

IMPROVED BURIAL-CASE.

Specification forming part of Letters Patent No. 81,437, dated August 25, 1868.

To all whom it may concern:

Be it known that I, FRANZ VESTER, of Newark, in the county of Essex, and State of New Jersey, have invented a new and Improved Improvement in Burial-Cases or Coffins for the Dead; and do hereby declare that the following is a full, clear, and exact description of the same, reference being had to the accompanying drawings, making part of this specification, of which—

Figure 1 is a top view; Fig. 2, a side elevation; Fig. 3, an under side view of the lid; Fig. 4, a longi

The nature of this invention consists in placing on the lid of the coffin, and directly over the face of the body laid therein, a square tube, which extends from the coffin up through and over the surface of the grave, said tube containing a ladder and a cord, one end of said cord being placed in the hand of the person laid in the coffin, and the other end of said cord being attached to a bell on the top of the square tube, so that, should a person be interred ere life is extinct, he can, on recovery to consciousness, ascend from the grave by the coffin by the ladder; or, if not able to ascend by said ladder, ring the bell, thereby giving an alarm, and thus save himself from premature burial and death: and if, on inspection, life is extinct, the tube is withdrawn, the sliding door closed, and the tube used for a similar purpose.

In the said drawings, A denotes the body of the coffin; B, the lid. C represents a square tube, which is seated in a square base, D, at the head to the lid of the coffin, and held in place by a spring-bar, E, connected with the sliding glass door L. This square tube C extends from the lid of the coffin to and above the surface of the grave, and has air-inlet openings F F, which communicate with the head of the coffin, and has also a glass door, near its top, which may be easily raised or looked through, for inspection of the person laid in the coffin. The said tube contains a

ladder, H, by which the person laid in the coffin may, on returning life, ascend to the surface of the earth; and the said tube has near its top a bell, I, from which a cord, K, is suspended, the lower end of said cord being placed in the hand of the person laid in the coffin, as shown in the drawings. On the outer side of the coffin-lid is a sliding glass door, L, held by a spring, M, which closes the coffin, excluding the air when the tube C is withdrawn from the coffin.

The operation of my invention is as follows: The person being laid in the coffin, A of the coffin, and the cord K placed in the hand of the corpse, the cord is next drawn through the tube C and attached to the bell I, and the tube C placed in the base D, on the lid of the coffin. The coffin is now lowered into the grave, and the grave filled up to the air-inlets F F. Now, should the person laid in the coffin, on returning life, desire to ascend from the coffin and the grave to the surface, he can do so by means of the ladder; but, if too weak to ascend by the ladder, can pull the cord in his hand, and ring the bell I, giving the desired alarm for help, and thus save himself from premature death by being buried alive. Should life be extinct, the tube C is removed, the door L closed, and the tube used for a similar purpose.

Having described my invention, what I claim as new, and desire to secure by Letters Patent, is—

1. The application of the tube C and ladder H to a burial-case or coffin, substantially and for the purposes described and set forth.

2. In combination with the tube C and ladder H, the cord K and bell I, for the purposes substantially as set forth and described.

In testimony whereof I have hereunto set my signature this 9th day of July, 1868.

FRANZ VESTER.

Witnesses:
A. NEILL,
R. SANGMEISTER.

AIRPLANE

Patent Name: Flying-Machine
Patent Number: 821,393
Patent Date: May 22, 1906
Inventor: Orville Wright and Wilbur Wright, of Dayton, Ohio

"Man must rise above the Earth—to the top of the atmosphere and beyond—for only thus will he fully understand the world in which he lives."—Socrates

What It Does Enables human beings to travel by means of flight.

Background On December 17, 1903, a telegram came from Kitty Hawk, North Carolina, that ushered in an important new chapter in history. Orville Wright informed his father of the success that he and his brother Wilbur had enjoyed that very morning when they tested their invention: the flying machine. On the last test of the morning, Wilbur had stayed airborne for fifty-nine seconds and traveled 852 feet.

Orville Wright

Wilbur Wright

To fly has been a dream of mankind since its beginnings; what took place that morning was, to a great extent, the culmination of exacting human will over countless eons. The Wright brothers had selected the outer banks of North Carolina for their windy atmosphere to help takeoff, and for the sands to soften their landing.

The Wright brothers' flying machine resembled a sophisticated version of a modern-day hang glider. Fastened into position, the operator propelled the apparatus forward along a rail; the biplane lifted off the ground propelled by both motor and wind. Its light-weight material and aerodynamic design were central to its success.

How It Works

The year 2003 marked a milestone for the airplane: the centennial of flight. On December 17, exactly a century after the Wright brothers succeeded, a reenactment was attempted—with unfortunate results. According to an Associated Press report, "a cheer rose from the crowd of 35,000 when the muslin-winged flyer roared to life and began moving down a wooden launch track. But that cheer suddenly turned to a groan when the rickety craft stopped dead in a muddy puddle at the end of the track."

The biplane design features two "aeroplanes"—parallel, superposed wings (1) and (2) in patent illustration figure 1—angled to such a degree as to be both propelled at a reasonable pace and lifted by wind. The aeroplanes are made of lightweight, sturdy, and flexible cloth-covered frames.

1. A series of transverse spars, longitudinal ribs, and diagonal threads forms a truss system to secure the planes together and provide a central component of the design.

2. V-shaped arms (23) help support a rear vertical rudder (22) attached by pivots (25), upon one of which is mounted a sheave or pulley (26).

3. Through the pulley, a tiller rope (27) passes, extending out laterally.

4. The tiller rope is secured to another rope (19) that passes along the top surface of the lower aeroplane, where shifting of the cradle (18) serves to turn the rudder.

5. At the front of the airplane, two struts (29) hold in place a horizontal rudder (31), the rear edge of which lies immediately in front of the operator and can be manipulated up or down.

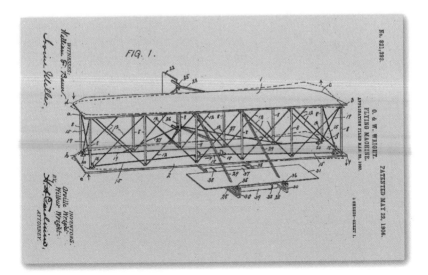

In the Inventors' Words

With a cruising speed of 1,350 mph, the Concorde is the world's fastest nonmilitary aircraft. It flies at supersonic speed on the edges of space, some 60,000 feet above ground. A trip from New York to London takes just over three hours; less than half the flying time for a Boeing 747. Business trends, and the events of September 11, 2001, have reduced market travel demand for this incredible aircraft; the Concorde dipped its nose for its farewell landing in late 2003.

"Our invention relates to that class of flying-machines in which the weight is sustained by the reactions resulting when one or more aeroplanes are moved through the air edgewise at a small angle of incidence, either by the application of mechanical power or by the utilization of the force of gravity.

"The objects of our invention are to provide means for maintaining or restoring the equilibrium or lateral balance of the apparatus, to provide means for guiding the machine both vertically and horizontally, and to provide a structure combining lightness, strength, convenience of construction, and certain other advantages.

"The apparatus is supported in the air by reason of the contact between the air and the under surface of one of more aeroplanes, the contact-surface being presented at a small angle of incidence to the air."

ARTIFICIAL HEART

Patent Name: Soft Shell Mushroom Shaped Heart
Patent Number: 3,641,591
Patent Date: February 15, 1972
Inventor: Willem J. Kolff, of Salt Lake City, Utah

What It Does

Assists or replaces the functions of a human heart.

Background

That the heart can be replaced successfully by an artificial substitute, even if only temporarily, is no less than a miraculous testament to human ingenuity. The blood vessels in an average human adult, if laid out in a single line, would wrap around the globe a few times. The heart is responsible for the distribution of blood through this sophisticated complex of veins and arteries. This is a big job that keeps getting bigger as we grow; and often more difficult if we risk heart failure from a variety of conditions including chronic high blood pressure, coronary artery disease, diseased heart valves, or cardiomyopathy.

As doctors learn more about the heart's intricacies, the artificial heart continues to evolve. But the original brain behind the heart belonged to the Dutch-born

American doctor Willem J. Kolff, a surgery professor at the University of Utah. Kolff received a patent for the first artificial heart on February 15, 1972. It happened to be the day after St. Valentine's Day, which happens to be Dr. Kolff's birthday: he was born on February 14, 1911.

How It Works

The artificial heart is a sophisticated device that works in two general cycles: filling and releasing.

In Figure 1:

1. A blood pumping chamber (10) is enclosed within a flexible wall (11).

2. A nonelastic, flexible net (12) restrains the wall from overdistending.

3. Blood enters the pumping chamber through an opening between a valve seat (15) and a valve head (14), atop the mushroom-shaped pumping member (13).

4. Another valve (16) prevents reverse flow of blood during the filling phase.

In Figure 2:

1. Air is forced into the mushroom-shaped pumping member.

2. While the pumping member expands, valve head (14) and valve seat (15) seal the chamber at the top, thereby preventing the backflow of blood.

3. Blood is then pumped out through the release valve (16).

In the Inventor's Words

"… It is desirable for the output volume of the artificial heart to be responsive to and substantially directly

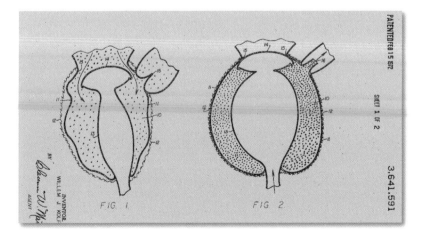

PATENTED FEB 15 1972

SHEET 1 OF 2

3,641,591

FIG. 1

FIG. 2

BY
WILLEM J. KOLFF
INVENTOR

AGENT

AbioCor Implantable
Replacement Heart

proportional to the blood input pressure…. The relationship between the output volume and inlet pressure is known as Starling's Law and is one of the important considerations in the design of an artificial heart, since this relationship will prevent the collapse of inlet blood vessels from an over suction by the pumping chamber or blood pooling from insufficient output volume."

CAMERA

Patent Name: Method of Taking Likenesses by Means of a Concave Reflector and Plates so Prepared as That Luminous or Other Rays Will Act Thereon
Patent Number: 1,582
Patent Date: May 8, 1840
Inventor: Alexander S. Wolcott, of New York, New York

What It Does   Captures live, still images on film, which can then be printed on paper and preserved as "photographs."

Background   In 1840, Alexander S. Wolcott became the first person in the United States to receive a patent for a photographic invention. In Europe, Frenchman Louis-Jacques-Mandé Daguerre had already developed a process using light to capture images onto plates. But these eponymous "daguerreotypes" required extremely long exposure times, so were best suited to inanimate objects. In 1839, Daguerre published a report on his process for producing daguerreotypes, and such creative and entrepreneurial Americans as Samuel F. B. Morse, who had been a portrait painter, and Alexander Wolcott seized upon the concept and set out to improve it. Wolcott, a dentist, came up with the idea of adding a concave mirror inside the camera to reflect intense light onto a plate or film, significantly speeding up exposure time. Wolcott and a partner opened the world's first photo portrait studio in New York just two months before this patent was issued.

How It Works

1. A small open steel frame (C) stands on a support (D) within a box.

2. The support is fixed to a piece of wood (E) or other material that can slide along the bottom of the box.

3. A door at the top of the camera (A) serves as an eyepiece.

4. A concave reflector placed inside the box should reflect the image as it is desired to appear through the small steel frame.

5. The material on which the likeness will imprint is placed in a small steel frame.

6. By action of the light-reflecting mirror against the material, the image of the person whose likeness is being taken becomes imprinted on the material.

In the Inventor's Words

"When the camera—that is, the box A with the reflector, &c.—is to be used, the person whose likeness is

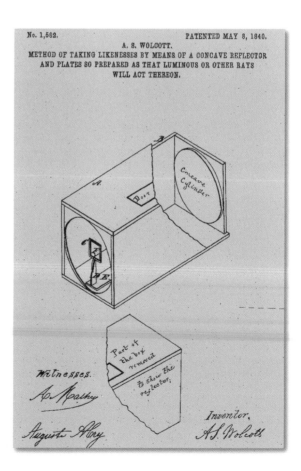

No. 1,582.   PATENTED MAY 8, 1840.

A. S. WOLCOTT.

METHOD OF TAKING LIKENESSES BY MEANS OF A CONCAVE REFLECTOR AND PLATES SO PREPARED AS THAT LUMINOUS OR OTHER RAYS WILL ACT THEREON.

to be taken should be placed in a chair, to which some suitable support for the head is attached to enable him to remain perfectly still. The camera should then be placed with the open end immediately opposite to the person. A trial-plate is then to be placed or put against the frame C and the focus adjusted by sliding the piece E. The trial plate is then to be removed and the plate, paper, or other material to be used (prepared in any of the well-known methods for being acted on by luminous or other ways) put into its place and allowed to remain as long as required to form the image."

DNA FINGERPRINTING

Patent Name: Method of Characterizing Genomic DNA
Patent Number: 5,175,082
Patent Date: December 29, 1992
Inventor: Alec J. Jeffreys, of Leicester, United Kingdom

What It Does

A person's DNA pattern is as distinctive as a fingerprint. Being able to create a genetic "fingerprint" from a DNA sample provides a definitive means of identification that can be used for paternity and maternity testing, in criminal investigations, or in forensic medicine.

Background

Sir Alec Jeffreys invented a genetic technique that has revolutionized criminal investigation and forensic science. DNA, or genetic fingerprinting, involves a technique used to distinguish individuals within a given species based on their unique DNA.

In September 1984, Jeffreys, a scientist and professor at the University of Leicester, England, was studying the evolution of genes at the British Antarctic Survey Headquarters. To track and mark the positions of genes, it was useful to isolate small areas of DNA where high levels of variability were found. Jeffreys was using a lump of seal meat as his subject, comparing the seal gene with the closest counterpart gene in a human. His comparisons led him to devise a method of detecting lots of variations in genetic material simultaneously, leading in turn to a surprising revelation.

In 1992, DNA fingerprinting confirmed that the remains of a body found in Brazil in 1979 belonged to Dr. Josef Mengele, the notorious "Angel of Death" of the Auschwitz death camp.

The DNA sequences in his genetic studies soon came to reveal signatures far more specific than those which distinguish species.

Jeffreys and his colleagues had hit upon the first DNA "fingerprint," and they knew right away that its implications were huge. A foolproof system of identification for establishing family relationships, investigating crime, conducting medical research, and even studying wildlife could have far-reaching applications. In 1994, the science that Jeffreys helped spawn made a very public appearance: prosecutors obtained and presented DNA evidence that put O. J. Simpson at the scene of a double murder. Now that DNA evidence is routinely used in criminal investigations, a number of people serving time for capital crimes have been exonerated and freed because the DNA evidence pointed to someone else.

Conventional fingerprinting preceded the more exacting DNA tests that now aid in identification.

How It Works

Every cell in the human body contains a copy of a DNA blueprint made up of twenty-three pairs of chromosomes. Only a tiny portion of the human genome varies from one individual to another. Methods of DNA fingerprinting focus on these small chromosomal regions of variability. Here DNA fragments reveal paired chromosomal sequences—variable number tandem repeats, or "microsatellites"—that repeat anywhere from one to thirty times in a row. Two different numbers of repeats in a single sequence come from the paired paternal and maternal chromosomes. These numbers, probed at various variable number tandem repeat locations, create a statistically unique DNA profile.

In the Inventor's Words

"The present invention relates generally to polynucleotides and DNA and RNA probes, their preparation and their use in genetic characterization. Such uses may include, for example, establishing human, animal or plant origin, and the polynucleotides and probes of the invention may thus find use for example in paternity disputes or forensic medicine or in the prevention, diagnosis, and treatment of genetic disorders or predispositions.

"In British patent application No. 8525252 (publication no. 2166445) there are described various DNA sequences which may be used as probes to hybridize individually at a number of polymorphic sites within the human and animal genomes enabling the production of a "fingerprint" composed of marked bands of differing molecular weights. The fingerprint as a whole is characteristic of the individual concerned and the origin of the differing bands can be traced through the ancestry of the individual and can in certain cases be postulated as associated with certain genetic disorders."

DYNAMITE

Patent Name: Improved Explosive Compound
Patent Number: 78,317
Patent Date: May 26, 1868
Inventor: Alfred Nobel, of Stockholm, Sweden

What It Does

This relatively safe and effective explosive (contained in rods, stable to transport, and able to be detonated from a distance) can create controlled, focused explosions of various intensities, for the purpose of demolition.

Background

Alfred Nobel was both a scientist and a worldly entrepreneur. The son of a wealthy inventor and engineer of bridges, he understood how a little ingenuity could be applied to turn a large profit and change the way work is done. To these ends, he came up with dynamite. One of the most important tools of the era, dynamite instantly made mining, construction, and highway and railroad building faster and cheaper; and it contributed significantly to the Nobel family fortune.

Before his invention was patented, Nobel worked with his father to develop nitroglycerine as a commercially viable explosive for blasting rocks. Experimenting with nitroglycerine, an oily fluid, proved a dangerous business that did not come without human cost. One early explosion killed several people, including the inventor's brother. But Nobel remained determined to succeed, and tinkered with organic additives in order to make the handling of nitroglycerine safer. By mixing silica with nitroglycerine, he found that the liquid could be made into a malleable paste. The paste could then be shaped into rods. The rods could be inserted into holes drilled where explosions would be safest and most strategic.

During his lifetime, Nobel held more than 350 patents in different countries in a variety of fields, including electrochemistry and physiology. He willed much of his fortune to establish a fund that awards achievements in Physics, Chemistry, Physiology and Medicine, Literature, and Peace. The annual Nobel Prize ceremony is held in Stockholm, Sweden, where Nobel was born.

How It Works

The chemical explosive nitroglycerine was discovered by Italian chemist Ascanio Sobrero in 1846.

1. Nitroglycerine and absorbent silica are mixed together at a ratio of approximately 60/40 for a minimally effective explosion, up to about 78/22 for a more powerful blast.

2. Confined in a tight, strong enclosure, the compound is detonated by the application of heat above the temperature of 360° Fahrenheit.

3. Ignition is accomplished by the use of a fuse or detonator, which can be activated from afar.

In the Inventor's Words

"So great is the absorbent quality of this earth, that it will take up about three times its own weight of nitroglycerine and still retain its powder-form, thus leaving the nitroglycerine so compact and concentrated as to have very nearly its original explosive power; whereas, if another substance, having a less absorbent capacity, is used, a correspondingly less proportion of nitroglycerine will be absorbed, and the powder be correspondingly weak or wholly inexplosive."

A miner places dynamite for demolition.

HELICOPTER

Patent Name: Direct-Lift Aircraft
Patent Number: 1,994,488
Patent Date: March 19, 1935
Inventor: Igor I. Sikorsky, of Nichols, Connecticut

What It Does

This compact, versatile, rotor-powered flying machine is capable of taking off and landing without the need for a runway, moving in multiple directions at variable speeds, flying close to the ground, and carrying passengers and cargo for relatively long distances.

Background

Igor Ivan Sikorsky was one of those individuals who seemed to find his path through life at an early age. Born in Kiev, Russia, he was fascinated by flight as a child; in his early twenties, he was a leading figure in Russian aeronautics. In 1913, when he was just twenty-four, Sikorsky built the world's first four-engine airplane, a close ancestor to the bomber airplanes of World War I.

Unlike airplanes, helicopters can fly backward, hover in mid air, rotate, and require only a minimum of space for take off and landing.

Pursuing his dreams, Sikorsky moved to the United States and established the Sikorsky Aero Engineering Corporation. From the mid 1920s to 1940, Sikorsky built a variety of large passenger-carrying amphibious aircraft. His "flying boats" garnered him success and fame, and, more importantly to him, enormous credibility in the realm of U.S. aeronautics.

All along, Sikorsky had been interested in creating a successful helicopter, and worked on the invention of a direct-lift single-rotor aircraft. Back in Russia when he was twenty, he had tried to build one, to no avail. But thirty years later, his determination bore fruit: On September 14, 1939, the VS-300 made its first flight. The first commercially successful helicopter has since become a prototype for single-rotor helicopters, the world's most versatile aircraft, and Sikorsky is viewed as one of history's most important contributors to the field of aeronautics.

1. A single rotary propeller at the top of the helicopter creates the necessary lift and can be tilted to maneuver the vehicle.

2. A smaller rotary propeller at the tail offers a steering mechanism.

3. While it seems relatively simple, operating a helicopter requires a great deal of coordination, practice, and the use of both feet and both hands.

4. One hand control is set to operate the lateral direction of the vehicle, which tilts the main propeller.

5. Another hand control is set to operate the vehicle's vertical direction by adjusting the speed of the main propeller.

6. Two foot petals control the tail rotor, which acts as a rudder for the vehicle.

Although he did not invent it, Igor Sikorsky is considered the father of the helicopter. He was inducted into the Inventors Hall of Fame in 1987.

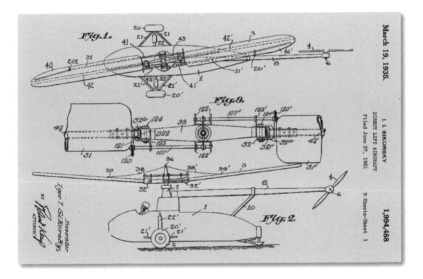

In the Inventor's Words

"The object of my invention is to solve successfully such direct lift aircraft problems as torque compensation, directional steering, and the application of power to a vertical lift producing propeller, either by a power plant or by means of air through which the vehicle navigates.

"A further object of the invention is the creation of a cheap, simple and easily operated mechanism for controlling the lateral and longitudinal stability of the

"The work of the individual still remains the spark that moves mankind ahead." —Igor Sikorsky

aircraft and guide it positively in its various directions of flight."

INTERNAL COMBUSTION ENGINE

Patent Name: Gas-Motor Engine
Patent Number: 365,701
Patent Date: June 28, 1887
Inventor: Nikolaus August Otto, of Deutz-on-the-Rhine, Germany

What It Does

Provides an apparatus for igniting gas-motor engines in which the charge is subject to compression before ignition. Otto's engine is the prototype of the combustion engine. The piston draws in and compresses a gas-air mixture inside a closed container. A spark ignites the mixture and it explodes.

Background

Fellow countryman and close colleague Gottlieb Daimler is believed to have used Otto's engine in building the world's first motorcycle in 1885. Daimler had been technical director at the company Otto co-owned and later went on to modify his own version of the gas engine.

The world's love affair with the automobile began as soon as the first Model T Fords started rolling down the streets, and it's been gaining momentum ever since. Yet, one of the most important innovations in automobile history was invented by someone who was never actually involved in making cars. Nikolaus Otto was a traveling salesman before he became interested in designing engines. He started an engine manufacturing company with a partner in 1864. In 1876, he devised the first practical four-stroke internal combustion engine.

Otto's company sold over 30,000 of the new engines over the next decade. Now called DEUTZ AG, KÖLN, the company Otto cofounded is the oldest manufacturer of internal combustion engines. The "Otto-Cycle Engine," as it was named in honor of its inventor, remains a model for engines made today. Indeed, on the long road of automobile history, the internal combustion engine was a major mile-marker, if not a true starting point.

How It Works

A piston is employed to facilitate the following four strokes of an internal combustion engine:

1. An intake valve opens to draw a mixture of air and gas into the cylinder.
2. The mixture is compressed in the cylinder.
3. A small gas charge explodes in the cylinder.
4. A valve opens to release the exhaust.

In the Inventor's Words

"This invention relates to an improved construction of apparatus for igniting the charges of those gas-motor engines in which the charge is subject to compression before ignition.

"The apparatus consists of a slide that is pressed against a slide-face of the cylinder by means of a loose cover acted on by springs, and is worked by suitable gearing from the engine-shaft, such slide being provided with posts and passages arranged and operat-

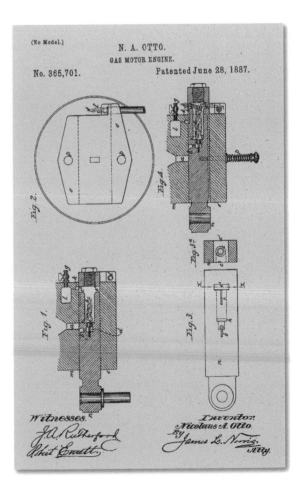

(No Model.)

N. A. OTTO.

GAS MOTOR ENGINE.

No. 365,701.

Patented June 28, 1887.

Witnesses.

Inventor
Nicolaus A. Otto.
By James L. Norris.
Atty.

ing as I will proceed to explain with reference to the accompanying drawing, in which—

"Figure 1 shows a horizontal section of the apparatus at the moment of igniting the cylinder-charge. Figure 2 shows an end elevation. Figure 3 shows a view of the inner face of the slide. Figure 3ᵃ shows a cross-section; and figure 4 shows the same view as figure 1, with the slide at its extreme right-hand position."

CHRONOLOGY OF THE AUTOMOBILE AGE

Perhaps no other invention has changed the lives of so many as the automobile. The prevalence and ease of use of the gasoline engine have given people a freedom of movement that has forever changed the face of the global landscape. But no one person can claim the honor of having invented the automobile as we know it today: The gasoline-powered car is the result of numerous innovations throughout history, beginning as far back as the eighteenth century.

1769–70: The first non-horse-drawn road vehicles were built by Nicolas-Joseph Cugnot, a French military engineer. He built a passenger vehicle and a cargo tractor, both of which were powered by steam. However, steam-powered vehicles were soon deemed too dangerous, dirty, and noisy for regular street use, and engineers started looking for new ways to power vehicles.

1860: Jean-Joseph-Étienne Lenoir, a Belgian-born Frenchman, patented the first viable internal-combustion engine. Similar to a steam engine, the single cylinder engine was fueled by streetlight gas and ignited by a storage battery, and was used mainly to power machines. But his invention proved too unwieldy for everyday use. After selling about five hundred engines, Lenoir died in poverty.

1885: Gottlieb Daimler and Karl Benz, both engineers in Germany, separately worked on developing engines based on principles of internal combustion. Daimler later went on to found the Daimler Motor Company in 1890, and unveiled the first Mercedes in 1901.

1887: German engine manufacturer Nikolaus Otto patented his "Gas-Motor Engine," a revolutionary invention in automobile history. It was the first commercially successful four-stroke internal combustion engine; it remains a prototype for today's engines.

1891: Long before Toyota introduced the Prius, American inventor William Morris unveiled the first electric car, powered by batteries that were stowed under the car's seats. Though quiet and nonpolluting, it required frequent recharging and couldn't go very fast. Its initial popularity soon died.

1893–94: Brothers Charles E. and J. Frank Duryea introduced the first gasoline-powered automobile in the U.S., and later founded the Duryea Motor Wagon Company, the first U.S. company to produce gas-powered automobiles.

1895: Michelin, a French company, introduced rubber tires filled with compressed air, giving automobiles their first smooth ride.

1896: Henry Ford built the first successful gasoline-powered car, an early prototype of the Model T.

1901: Huge oil fields were discovered in eastern Texas, making gasoline cheaper and more accessible, and helping accelerate the popularization of the automobile.

1901: Ransom E. Olds began mass-producing his namesake car, the Oldsmobile. It features a curved dashboard, injecting the automobile with a sense of style.

1903: Henry Ford founded the Ford Motor Company in Michigan.

1904: Henry M. Leland's Cadillac Automobile Company begins manufacturing cars with interchangeable parts, helping to streamline the production process.

1908: Henry Ford unveiled the iconic Model T, which comes to be known affectionately as the Tin Lizzie. The Model T outsold all other cars for the next twenty years.

1911: Charles F. Kettering invented the electric self-starter. This ingenious invention allowed drivers to start their cars electronically, rather than using the cumbersome hand cranks that were until then standard.

1913: Henry Ford's automobile factory installed a moving assembly line, increasing the speed at which cars could be manufactured, decreasing the cost of the final product, and making cars more affordable for the masses.

1914: U.S. automakers agreed to cross-licensing, the system of sharing patents without having to pay one another for the information. This collaborative spirit helped accelerate innovation in automobile manufacturing in the early part of the twentieth century—but inevitably, the system of cross-licensing ended in 1956.

1918: Malcolm Loughead invented the hydraulic four-wheel brake system, making it easier and safer for drivers to bring their cars to a halt.

1922: Low-pressure tires, known as balloon tires, were introduced. The decrease in pressure made riding in a car more comfortable.

1939: Technological improvements contributed to the ease of operating an automobile and driving long distances. The fully automatic transmission decreased physical exertion by eliminating the need to switch gears manually, while air-conditioning systems helped control the comfort of car's interior.

1956: The Federal-Aid Highway Act of 1956 was introduced under President Dwight D. Eisenhower. It earmarked federal funds for the completion of a nationwide network of highways in the United States.

1968: In order to help control environmental pollution, the U.S. mandated that automobiles be equipped with devices to reduce exhaust fumes.

1975: The worldwide oil shortage led the U.S. Congress to pass a law requiring that cars be made more fuel efficient.

1997 and beyond: Toyota introduced the Prius, a mass-produced hybrid car that combines the gas engine with an electric motor. Honda followed suit in 2000 with their hybrid car, the Insight. Auto manufacturers began exploring alternate sources of fuel energy, such as the fuel cell, which burns hydrogen as fuel, producing only water as a by-product. The car of the future will certainly be less polluting and more fuel-efficient than ever before.

LIGHTBULB

Patent Name: Electric-Lamp
Patent Number: 223,898
Patent Date: January 27, 1880
Inventor: Thomas A. Edison, of Menlo Park, New Jersey

What It Does Provides light by means of a glowing filament inside a glass globe, powered by electricity.

Background Thomas Edison was a genius credited with a great number of inventions, but the electric lightbulb was not entirely his own creation. Many inventors throughout the world were trying to replace the gas lamp with some sort of lightbulb. In fact, Edison's British contemporary Joseph Swan beat him to the punch when Swan demonstrated the world's first electric lightbulb in 1878. A few men before this—the Canadian Henry Woodward and his partner Matthew Evans—patented a lightbulb, but they didn't have the capital necessary to commercialize their invention. Edison, who was well established by then, purchased the rights to their patent.

Young Thomas Edison

But Edison was not simply buying his way into another invention. He worked at breakneck speed to invent a variety of components that would ensure that his light would shine with the brightest commercial success. Ultimately, Edison ran his new lamp for two days and forty minutes. The day that light burned out is the date usually given for the first commercially practical lamp—October 21, 1879. By 1880, Edison had perfected a 16-watt incandescent bulb that could burn for up to 1,500 hours.

How It Works 1. Glass is blown into a bulb-shaped mold.
2. A thin wire or filament is drawn out and wound into a double coil.
3. At both ends, the filament is attached to power leads embedded in a supporting glass structure.

On September 4, 1882, the first commercial power station began providing light and electricity to customers in a one-square-mile area of lower Manhattan.

Incandescent light-bulbs—those which emit light as a result of heat—are less efficient than more recent innovations: neon, fluorescent lighting, and LEDs (light emitting diodes). Conventional lightbulbs use significantly more energy heating the filament, wasting electricity in the process. A 60-watt incandescent lightbulb lasts about 750 hours, about ten times less than a compact fluorescent bulb that generates the same amount of light.

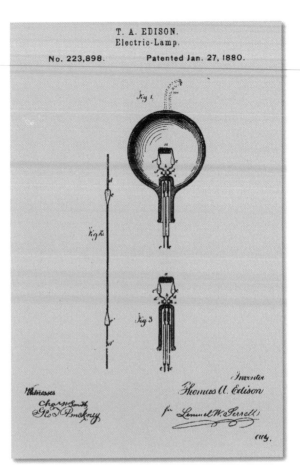

T. A. EDISON.
Electric-Lamp.

No. 223,898.   Patented Jan. 27, 1880.

4. This support is inserted into a bulb and the two glass parts fused together.

5. A tube in the support is used to replace the air inside of the bulb with inert gases.

6. The tube is cut, the opening is sealed, and the base of the bulb is attached.

7. Electric currents flow from one contact to the other, traveling through the wires and heating the tungsten filament, which in turn emits light.

In the Inventor's Words

"Be it known that I, THOMAS ALVA EDISON, of Menlo Park, In the State of New Jersey, United States of America, have invented an Improvement In Electric Lamps, and in the method of manufacturing the same (Cane No. z 188,) of which the following is a specification.

"The object of this invention is to produce electric lamps giving off light by Incandescence, which lamps shall have high resistance, so as to allow the practical subdivision of the electric light."

INVENTOR PROFILE

Thomas A. Edison
(1847–1931)

No book about patents would be complete without a profile of Thomas Alva Edison. Edison was awarded 1,093 patents—the greatest number of patents the U.S. Patent Office has ever given to a single person. In addition, he received thousands more patents from dozens of other countries. His inventions helped advance the technology of electric lighting, phonographs, telegraphs, telephones, movies, and more. He also established one of the first modern research laboratories, which some scientists feel is the most important contribution of his vast legacy.

Born in 1847 in Milan, Ohio, Edison was the youngest of seven children. Although he was an industrious and clever child, it could well have been a serendipitous meeting that led Edison into a life of inventing. At the age of fifteen, Edison happened to be on hand to save the son of a telegraph operator from an oncoming railroad car. To thank him, the boy's father gave Edison free lessons in operating a telegraph. This led to a job with the Western Union Telegraph Company, which in turn awoke in Edison a lifelong interest in mechanics, electronics, and chemistry.

In 1868, Edison moved to Boston to continue working as a telegraph operator. There he displayed some of the first hints of his knack for innovation: He came up with a plan to improve a town fire-alarm system, and enhanced the printed outputs of telegraphs for the stockbrokers who relied on them for business. The next year Edison moved to New York City, where he forged more friendships with prominent leaders in finance and industry. These connections would later prove pivotal in helping him launch the many business ventures tied to his inventions. For example, in

1870, Edison moved to Newark, New Jersey, and co-founded a stock ticker manufacturing company based on technology he helped develop. But it wasn't until 1876—when Edison established his famous Menlo Park laboratory—that the inventions started coming fast and furious.

Through his years of work with the telegraph, Edison had become an expert, and often fiddled with the new telephone technology, patented by Alexander Graham Bell in 1876, that allowed telegraphs to "speak." In one of his first significant inventions, Edison fine-tuned the telephone transmitter so that sound came through louder and clearer, making talking on the telephone easier and more practical. Next, in 1877, Edison unveiled his most spectacular invention to date: the phonograph. The ability to record and play back voices, sounds, and music seemed almost magical at the time, and Edison garnered fame and recognition around the world.

As if that weren't enough, the very next year, Edison came up with one of his brightest ideas yet: the electric light. Inventors around the world were already experimenting with the carbon arc light, which produced a bright, intense light for a short period of time. Edison bested them, though, by creating the incandescent lamp, which passed electricity through a wire filament, causing the filament to glow. This type of lighting was safer, longer-lasting, and more practical for home use than the carbon arc. In 1879, Edison and his associates figured out how to make a carbon filament out of burnt sewing thread, and the precursor to the modern lightbulb was born. By 1880, Edison had perfected a 16-watt bulb that could burn for up to 1,500 hours. He also led the charge to create an industry to supply homes and businesses with the electricity necessary to power lightbulbs. As a result, he and other industry leaders banded together to form the General Electric Company in 1892.

Many other inventions and improvements upon existing technologies followed. Edison helped found the motion picture industry, established an iron ore processing plant, designed and improved batteries, and ventured into the cement business among other forays. He was certainly one of the great scientific geniuses of our time but he modestly refuted the notion that he was gifted and credited his success to hard work. As he was fond of saying, "Genius is one percent inspiration and ninety-nine percent perspiration."

NUCLEAR REACTOR

Patent Name: Neutronic Reactor
Patent Number: 2,708,656
Patent Date: May 17, 1955
Inventors: Enrico Fermi, of Santa Fe, New Mexico, and Leo Szilard, of Chicago, Illinois.

"I know not with what weapons World War III will be fought, but World War IV will be fought with sticks and stones."—Albert Einstein

What It Does
This device initiates and controls a self-sustaining nuclear chain reaction, and the massive amount of energy produced is typically used for power generation—but it has also been used as a weapon of mass destruction. The neutrons and fission byproducts are used for various military, experimental, and medical purposes.

Background
In 1938, German scientists Otto Hahn, Lise Meitner, and Fritz Strassmann discovered that by bombarding the nucleus of a uranium atom with thermal neutrons, the atom would split. In splitting one uranium atom, additional neutrons spontaneously formed that could then be used to split additional uranium atoms. The potential chain reaction was a mind-boggling discovery. As it splits into fragments, a heavy nucleus like the isotope of uranium can release several hundred million electron volts of energy. This discovery, that mass

The worst nuclear power accident occurred in 1986 in what is now Ukraine, in the former USSR. One of four reactors at the Chernobyl nuclear power plant lost control of a chain reaction, which caused explosions that blew the steel and concrete lid off of the reactor, unleashing massive amounts of radiation. More than 30 people died and 135,000 people were evacuated.

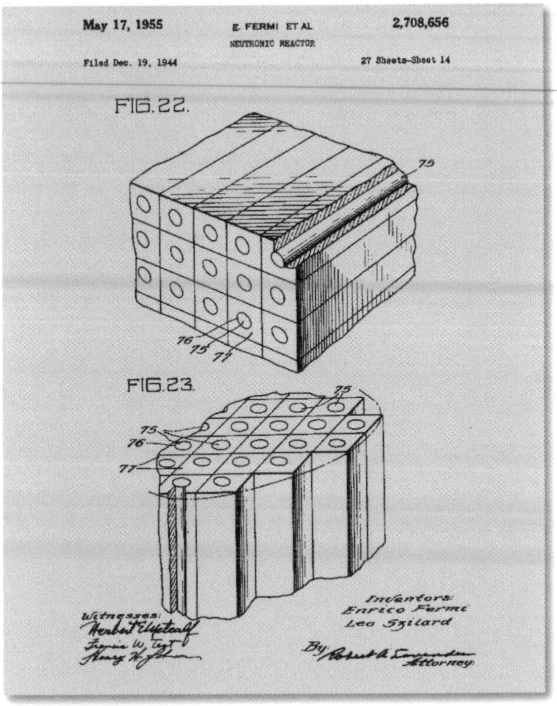

On August 6, 1945, just three weeks after the July test in New Mexico, an atomic bomb was dropped on Hiroshima, Japan. Tens of thousands were instantly killed; 65 percent of the city's structures were damaged. Including radiation deaths within a year, the bombing killed more than 155,000 people and effectively ended World War II.

is convertible to energy, jump-started a race among nations to build the atomic bomb.

From 1939 to 1945, more than $2 billion was spent on what was then called the Manhattan Project. Directed by Robert Oppenheimer in Los Alamos, New Mexico, the project was a top-secret U.S. effort that involved scores of brilliant scientists, including Italian physicist and Nobel Prize recipient Enrico Fermi and his Hungarian colleague Leo Szilard, who worked out of a secret lab in Chicago. There were many challenges: The desired uranium-235 was difficult to extract, and it was nearly identical to uranium-238. As isotopes, their chemical compositions are similar, and ordinary extraction methods could not separate them efficiently. The scientists needed to develop a mechanical means to separate them, as well as a method for producing a self-sustaining nuclear chain reaction. Fermi and Szilard filed a patent for their invention in 1944, but it wasn't approved until 1955, six months after Fermi's death.

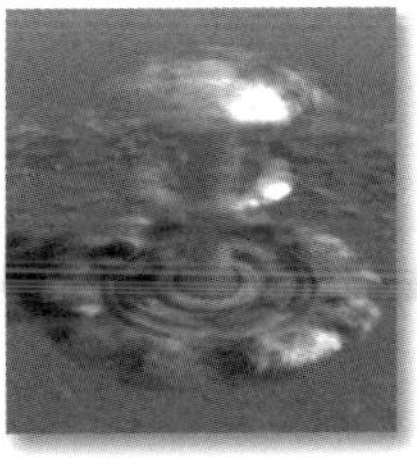

An atomic explosion.

Although the original intention was to create weapons of mass destruction, today's nuclear reactors are viable power sources that provide 17 percent of the world's electricity. There are some four hundred nuclear power plants in the world, with about one hundred on U.S. soil.

How It Works

A nuclear reactor enables a large but finite number of subsequent nuclear fissions to take place just below critical mass. Typically, such reactors are based on some variation of the following model:

1. Bundles of uranium rods that act as extremely high energy heat sources are submerged in water.
2. Control rods that absorb neutrons are inserted into the uranium bundles.
3. Raising the control rods causes the uranium core to produce more heat; lowering them causes them to produce less.
4. The water becomes steam, which drives a turbine, which, in turn, spins a generator.
5. Housed inside a radiation-proof concrete liner is the reactor's pressure vessel, where the reactor core and other key operating components are located.

In the Inventors' Words

"We have discovered certain essential principles required for the successful construction and operation of self-sustaining neutron chain reacting systems (known as neutronic reactors) with the production of power in the form of heat. These principles have been confirmed with the aid of measurements made in accordance with the means and method set forth in the above-identified application, and neutronic reactors have been constructed and operated at various power outputs, in accordance with these principles, as will be more fully brought out hereinafter."

Enrico Fermi

1896: Antoine Henri Becquerel conducted research on radiation, inspiring experiments two years later by Marie Curie.

1898: Marie Curie, a French physicist, working with her husband Pierre, discovered the radioactive elements radium and polonium.

1905: Albert Einstein published three papers that astounded the scientific community and established three new branches of physics. One of the three papers confirmed the atomic theory of matter.

1911: Georg von Hevesy came up with the idea of using radioactive material as a "tracer." The principle was later applied to medical diagnostic procedures.

1938: German scientists Otto Hahn, Fritz Strassmann, and Lise Meitner demonstrated nuclear fission, showing the scientific community how a uranium atom can be split by bombarding it with neutrons. As the nucleus of the uranium atom is split, it loses mass, which is then converted into energy. The process can also start a chain reaction as other neutrons are formed, bombarding the nuclei of other atoms. The implications of this process on weapons development will soon be exploited.

1939: Albert Einstein wrote a letter to President Franklin D. Roosevelt, warning him that it was possible to build an extremely powerful bomb using atomic energy, and that Germany might already be in the process of doing so. Einstein's letter prompted the U.S. to begin an investigation into the development of the atomic bomb.

December 1941: Japan bombed Pearl Harbor, inciting the U.S. to enter World War II.

September 1942: The U.S. formed the Manhattan Project in a race to develop the atomic bomb before the Germans. Secret laboratories were soon established at various locations across the U.S.

November 1942: Robert Oppenheimer was named director of the atomic bomb laboratory in Los Alamos, New Mexico.

December 1942: Enrico Fermi and others at the University of Chicago achieved the first sustained nuclear chain reaction, heralding the dawn of the atomic age.

July 1945: Scientists from the Los Alamos laboratory detonated the first atomic device, which had a plutonium core and released as much energy as 18,600 tons of TNT. The test came to be known as the Trinity Test.

August 1945: The United States dropped two atomic bombs on Japan. On August 6, 1945, military forces dropped a uranium bomb on Hiroshima, killing 70,000 people and injuring as many. Three days later, the U.S. dropped a second atomic bomb on Nagasaki, killing 40,000 and injuring 60,000. Japan surrendered, bringing an end to World War II.

March 1946: Winston Churchill declared the existence of an "iron curtain" in Europe, created by the Soviet Union. By the end of the decade, U.S.-Soviet tensions escalated into what became known as the Cold War.

July 1946: The U.S. conducted nuclear bomb tests on the Bikini Atoll in the Pacific.

1949: The Soviet Union tested its first atomic bomb.

1950: President Harry S. Truman ordered the Atomic Energy Commission to investigate the possibility of building the H-bomb, or hydrogen bomb.

1951: Usable electricity was produced from nuclear energy at the National Reactor Testing Station in Arco, Idaho.

1954: The Atomic Energy Act of 1954 was passed, a measure that promotes the peaceful use of nuclear energy.

1957: The first full-scale nuclear power plant began operation in Shippingport, Pennsylvania.

1961: President John F. Kennedy urged Americans to build fallout shelters in case of nuclear war.

1963: The U.S. and the Soviet Union signed the Limited Test Ban Treaty, which prohibited nuclear tests underwater, in the atmosphere, and in space. Underground nuclear tests were not banned.

1968: The Nuclear Nonproliferation Treaty was signed, which barred countries from making or assisting other countries in building nuclear weapons. Countries that didn't already have nuclear weapons pledged to refrain from building them. By 1986, 130 countries had ratified the treaty.

1978: The United States dropped plans to develop the neutron bomb, a specialized type of hydrogen bomb that destroys all life but leaves buildings intact.

1979: The Three Mile Island nuclear power plant near Harrisburg, Pennsylvania, suffered a partial meltdown. Low levels of radioactive material were released.

1983: The Nuclear Waste Policy Act of 1982 was signed, initiating the development of a nuclear waste repository.

1986: The Chernobyl nuclear reactor in Russia went through a meltdown and burst into flames. Over 135,000 people living within an 18-mile (30 kilometer) radius of the reactor were evacuated. Massive amounts of radioactive material were released into the environment.

1990: The Cold War formally ended with the Conference on Security and Cooperation in Europe.

1993 and beyond: The Department of Energy turned its attention toward cleaning up nuclear contamination. Carefully controlled nuclear weapons arsenals still exist, but are considered deterrents to large-scale wars rather than signs of aggression.

PENICILLIN

Patent Name: Method for Production of Penicillin
Patent Number: 2,442,141
Patent Date: May 25, 1948
Inventor: Andrew J. Moyer, assignor to the USA as represented by the Secretary of Agriculture, Peoria, Illinois

What It Does　Provides a practical, commercially viable way to mass-produce penicillin for use as an antibiotic.

Background　Derived from penicillium mold, penicillin was discovered in 1928 by Sir Alexander Fleming, who observed that bread mold could kill colonies of bacterium. This observation, and the conclusions that Fleming drew from it, proved pivotal in the annals of medical history: Fleming had discovered that natural substances could be used to kill disease-causing bacteria in the human body. Penicillin would prove to be one of the most important antibiotics ever. The problem was in extracting the antibiotic from its source, the penicillium mold: it took too long to get too little.

Penicillium mold (enlarged here many times) is a fungus that differs little from one that appears on bread in warm, humid weather.

During World War II, the need for penicillin became pressing to treat a rising number of wounded.

In the late 1950s, MIT professor John Sheehan was able to successfully synthesize penicillin. Sheehan received dozens of patents during his life. He worked on a variety of inventions including a method of making explosives used in rockets and torpedo warheads.

Corn steep liquor is still used in the fermentation processes for making many antibiotics, including penicillin.

Struggling to find a way to produce it efficiently, scientists tried two general approaches. One approach was synthesizing the drug: Researchers worked to figure out the molecular structure of the antibiotic to duplicate it synthetically. The other approach was to speed up production of the natural antibiotic itself.

More than a thousand scientists attempted to uncover the mysterious molecular structure of penicillin to synthesize the molecule, to no avail. Meanwhile, international efforts concentrated on developing a means to mass-produce the natural antibiotic. Andrew Moyer, a microbiologist, led the research in the United States. In culturing the mold in lactose and corn steep liquor, Moyer discovered that he could successfully and dramatically expedite the production of the antibiotic. Although his technique was not patented until 1948, it saved thousands of lives during the war.

How It Works

The penicillium mold is cultured in lactose and corn steep liquor.

In the Inventor's Words

"This application is made under the act of March 3, 1883, as amended by the act of April 30, 1928, and the invention herein described, if patented, may be manufactured by or for the Government of the United States of America for governmental purposes without the payment to me of any royalty thereon."

"Penicillin is a bactericidal or bacteriostatic compound that may be produced by *Penicillium natatum* Westling, as reported by Fleming (Brit. Jour. Exper. Path. 10,226 [1929]), who first observed the inhibition of bacterial growth in the vicinity of *P. notatum* colonies of petri-dish cultures. Penicillin possesses extremely valuable antibacterial properties which favor its application in the treatment of numerous infections. It is especially effective in the repressing or killing gram-positive staphylococci, pneumococci, gonococci, and numerous streptococci."

"I have found a method of increasing the penicillin content of culture liquors of a penicillin-producing mold, such as *Penicillium notatum* Westling, *Penicillium chrysogenum* Thom, *Penicillium baculatum* Westling and *Penicillium cyaneo-fulvum* Biourge, many-fold beyond that taught by the prior art, and to such an extent that production of penicillin for use as a therapeutic agent is now more feasible on a commercial basis. In addition to giving a better yield of penicillin, the present invention offers numerous other advantages, which will be apparent from the following description."

REFRIGERATION

Patent Name: Refrigerating Apparatus
Patent Number: 1,886,339
Patent Date: November 1, 1932
Inventor: Charles F. Kettering, of Dayton, Ohio

What It Does   Provides a safer means of refrigerating perishables, to keep food safe, edible, and appetizing for longer periods.

Background   Refrigeration was around long before this patent was issued. Ancient cultures had devised means of keeping meat in cold storage places; ice boxes were used in England in the early 1800s; technology led to the

employment of gas compression to create heat-absorbing liquid gases. By the 1920s, some of the first electric gas-compressing refrigerators were already on the market. At the same time, leakage of some of the toxic gases being compressed as refrigerant was proving dangerous and in some instances fatal.

   In 1928, industrial magnate and inventor Charles Franklin Kettering, then working for the Frigidaire Corporation, worked on developing a safer refrigerant. The result: Freon. It spawned a new method of refrigeration that has since been widely adopted as a standard. This patent represents the first commercially successful method of refrigeration to employ the use of the new gas.

How It Works   1. Two generator-absorbers (10, 11) are charged with absorbent.
2. Absorbers connect by means of circuits to respective evaporator mechanisms.

3. When one generator-absorber is heated and the other is cooled simultaneously, liquid refrigerant is supplied to the evaporator.

4. The liquid is condensed and gaseous refrigerant is withdrawn from the evaporator.

In the Inventor's Words

"The means for heating either generator-absorber is a burner (22) which may be placed at the lower extremity of the flue provided by the tube (15). The means for cooling either generator-absorber is a flue with which is associated a fan or blower (24) and which may be placed to direct a current of air through the flue (15). As illustrated, the blower is of the centrifugal type having its intake at the connection (27), but the particular form of air moving device is immaterial to the invention."

Freon is a chlorofluorocarbon. CFCs were later discovered to cause damage to the ozone layer of our atmosphere, but are still in wide use today.

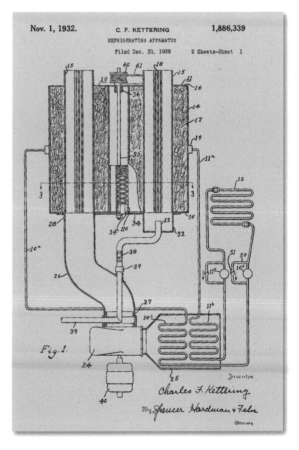

ROCKET

Patent Name: Rocket Apparatus
Patent Number: 1,102,653
Patent Date: July 7, 1914
Inventor: Robert H. Goddard of Worcester, Massachusetts

What It Does

This self-contained projectile propels itself into sustained flight through a series of combustions.

Background

Robert Goddard is the father of modern rocket science. While much of the world was still in awe of aircraft that could carry humans through the sky, Goddard was conducting experiments that would ultimately help carry humans far beyond our skies and into space.

Robert Goddard (second from left) poses with an early rocket.

This patent represents a first of its kind in illustrating the rocket propulsion theories for which Goddard became so famous. Just a year after he was awarded this patent, Goddard proved how thrust could be produced in a vacuum, bringing mankind one giant step closer to space flight. His expertise would first be applied to the military purposes at hand. During World War I, he developed solid-fuel rocket-launching weapons. This patent is for a rocket that uses solid fuels. He later developed liquid-fuel rockets that would be used to assist aircraft takeoff from naval carriers. He died in 1945, only a few years short of seeing the first rocket launched into space.

The apparatus described in this patent is a rocket specifically adapted to carry photographic devices. After the rocket finishes its flight, a second propelling charge sends the recording device into flight in an auxiliary rocket. Although not detailed in Goddard's patent, he mentioned that such devices could be safely returned to earth by means of parachute.

How It Works

1. Ignition or detonation of a rocket sets off a series of combustions that serve to propel the apparatus into a sustained flight.

2. In figure 1, a cross-section of the rocket reveals a combustion chamber (12) with a tapered tube (11) at the base.

3. Within the tube, a fuse (14) is ignited.

4. A series of rotating disks within the combustion chamber are set into motion through consecutive explosions.

5. Figure 3 depicts one disk that includes around its periphery a filament (18) that ignites the explosive charges arranged on the disk.

6. When the propelling charge in the main rocket is sufficiently exhausted, a second fuse ignites the smaller rocket.

7. The auxiliary rocket is launched out of firing tube 24.

8. Within the rocket head (29) a support is arranged to contain a recording instrument or camera.

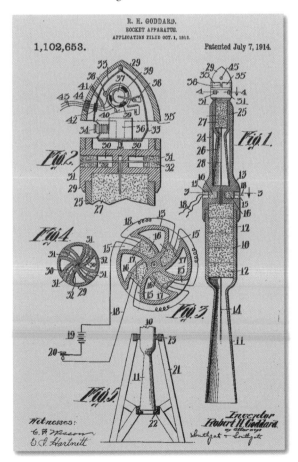

Robert Goddard

"Mathematical analysis shows that in any rocket apparatus of given mass, the necessary propelling charge varies according to an expression in which the percentage of the heat energy of the charge which is transformed to kinetic energy enters in an exponential relationship. Hence, any increase in the efficiency in the transformation results in greatly increased velocity of the apparatus and also permits a reduction in the amount of explosives used."

SKYSCRAPER STEEL

Patent Name: Improvement in the Manufacture of Iron and Steel
Patent Number: 49,051
Patent Date: July 25, 1865
Inventor: Henry Bessemer, of London, United Kingdom

What It Does

Provides a strong material for construction of tall buildings and other massive structures.

Background

George Fuller built one of New York's first skyscrapers, the Flatiron Building, in 1902. Fuller understood how to build upward, by resolving load-bearing problems in structural design. He created steel caging within exterior walls, and he used a new type of steel, created by Englishman Henry Bessemer.

The sky became the limit when Bessemer developed a radical new iron- and steel-making process—the first such process for mass-producing steel inexpensively. Bessemer used a furnace to repeatedly blast air into a container of fluid metal, thereby removing carbon. The new process allowed for a significantly stronger material and offered a seemingly limitless dimension to the world of architecture.

This patent sets forth the principles of Bessemer's original process and introduces an improved use of large-scale metal molding casts. Bessemer included a crucial new concept in molding sturdier steel. Previous molds for pole-shaped structures, for example, included two plate casts that were sealed shut by lugs and bolts, as airtight as possible, to be removed after

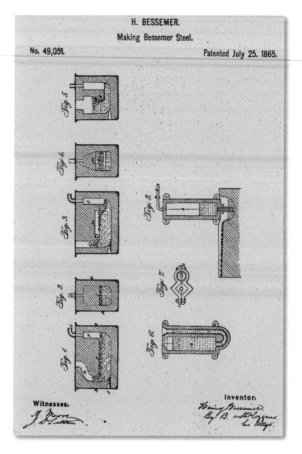

H. BESSEMER.
Making Bessemer Steel.

No. 49,051.

Patented July 25. 1865.

Fig. 5.

Fig. 4.

Fig. 3.

Fig. 8.

Fig. 7.

Fig. 2.

Fig. 1.

Fig. 6.

Witnesses.

Inventor.

The word "skyscraper" was coined in the 1880s, long before some of the tallest structures were built. In March 1996, the Petronas Towers in Kuala Lumpur added 241 foot-long steel pinnacles on top, in the process overtaking the Sears Tower as the world's tallest office building. The towers stand at 1,482 feet 6 inches.

casting. Air would inevitably become trapped inside, weakening the material. Bessemer realized that leaving an opening created a natural vacuum, making the steel stronger. Whether Bessemer knew it at the time or not, his new methods would soon dramatically alter the skylines of cities around the globe.

How It Works

1. A chamber made of brick, firestone, or another slow conductor of heat features a raised outlet.
2. The chamber is filled to capacity with fluid metal; no leakage is allowed from the opening.
3. Air or steam is forced into the metal by means of a pipe, displacing the liquid and allowing for the escape of air through the outlet.
4. A tapping hole sealed with loam is punctured to allow the refined, decarbonized metal to flow into suitable molds.

"My present invention relates, first, to peculiar modes by which the decarbonization or partial decarbonization and refinement of iron is effected by means of currents or jets of atmospheric air or steam, (alone or mixed,) which are made to impinge upon the surface or pass through or in contact with the metal while in a fluid state; secondly, in the manner in which the metal so treated is formed into ingots or masses suitable for being afterward made into bars, plates, or rods by the process of hammering or rolling."

SUBMARINE

Patent Name: Submarine Vessel
Patent Number: 581,213
Patent Date: April 20, 1897
Inventor: Simon Lake, of Pleasantville, New Jersey

What It Does

A submarine is an airtight and pressurized vessel for manned travel deep beneath the ocean's surface over a prolonged period. Submarines have been used in warfare as well as for various kinds of biological and oceanographic research.

Background

Like many kids of his day, Simon Lake read the science-fiction adventure stories of Jules Verne, and his life's work would be directly influenced by the experience. After reading *Twenty Thousand Leagues under the Sea*, Lake became determined to create a real-life "Nautilus," a state-of-the-art submarine.

The United States Navy has seventy-four armed submarines, more than any other navy.

In the late 1800s, Lake and John Holland were the two most prominent figures in designing the modern submarine. Their ideas were different, however.

A submarine surfaces in Lake Erie.

Holland's submarines dove and surfaced at an angle; Lake's used control pumps to manipulate the boat's buoyancy, allowing it to move vertically in place. With wheels, Lake's submarine could rove along the sea bottom as well. Both designs had their appeal. While Holland did win a contract from the Navy, it was Lake who later solved the problem of vision by fashioning the earliest-known periscope. Lake is also credited with the development of basic submarine technologies including even-keel hydroplanes, ballast tanks, and a twin-hull design.

While still in his twenties, Lake built his first experimental submarine, naming it the *Argonaut, Jr.* Encouraged by its success, he formed the Lake Submarine Company of New Jersey the following year, and built another submarine: the *Argonaut.* In 1898, this submarine was considered the first to demonstrate successful operation and long-distance navigation in the open sea.

Lake's achievement inspired a letter from Jules Verne, who wrote: "While my book *Twenty Thousand Leagues under the Sea* is entirely a work of imagination, my conviction is that all I said in it will come to pass. A thousand mile voyage in the Baltimore submarine boat (The Argonaut) is evidence of this. This conspicuous success of submarine navigation in the United States will push on under-water navigation all over the world. If such a successful test had come a few months earlier it might have played a great part in the war just closed. The next great war may be largely a contest between submarine boats."

In Figure 2:

How It Works 1. Pumps are arranged to fill and empty tanks (i, j).
2. An engine (e) is powered by furnace (f).
3. Within the hollow keel (M), a shaft (c') is rotated by the engine, turning both the screw propeller (p) at the aft and the front wheel (B).
4. Weights (s) arranged below the vessel are connected by water-tight cables wound around casings (g).
5. The submarine "driver" is stationed in a conning tower (I) with a steering wheel (K).
6. A companion hatch (O) is situated directly behind the conning tower.

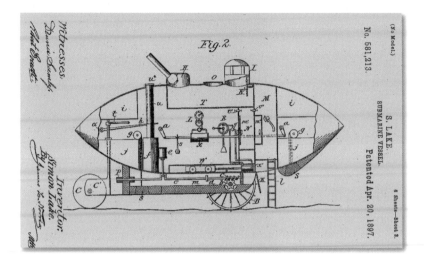

(No Model.)

No. 581,213.

S. LAKE.
SUBMARINE VESSEL.
Patented Apr. 20, 1897.

6 Sheets—Sheet 2.

Fig. 2.

Witnesses.

Inventor.
Simon Lake.
By

7. A diver's room (M) is situated directly in front and below the conning tower.

In Figure 1:

1. Extending from a turret (H) behind the companion hatch by a ball-and-socket joint is an observation tube (G).

2. Individual crew members are hoisted inside the tube to obtain an unobstructed view of the horizon through a rounded heavy glass top (g¹).

In the Inventor's Words

"My invention relates to an improved submarine vessel, and has for its object, first, to provide novel means for sinking the vessel to the bottom of the water when it is at a state of rest or has no headway and for permitting the vessel to rise to the surface of the water; second, to provide means whereby the vessel is enabled to travel upon the bottom or bed of the water; third, to provide a mechanism automatically controlled by the pressure of the water for submerging the vessel and maintaining it submerged at any desired or predetermined depth when under way; fourth, to provide means for automatically maintaining the vessel on a level keel irrespective of the disposal or shifting of the weights in the vessel; fifth, to provide novel means for affording ready ingress and egress from and to the vessel when submerged, and, lastly, to improve the construction generally and render more safe and certain the operation of submarine vessels.

TELEGRAPH

Patent Name: Improvement in the Mode of Communicating
Information by Signals by the Application of Electro-Magnetism
Patent Number: 1,647
Patent Date: June 20, 1840
Inventor: Samuel F. B. Morse of New York, New York

"What hath God wrought?" —Samuel Morse's first telegraph message

What It Does   Offers a system to relay audible messages by means of code over long distances via electromagnetic waves.

Background   If Samuel Finley Breese Morse was skeptical about the nature of technological advancement, his life belied the fact. He abandoned a successful career as a portrait painter to embark upon the improvement and creation of technology as an inventor. In the 1830s, he became interested in electronics and telegraphy. He envisioned a system of relaying messages in an electronic alphabet of dots and dashes through electromagnetic waves. His first successful system involved a tape on which a register printed his self-named Morse code language of dots and dashes. A telegraph operator would move a key on the device that would manipulate the circuit of electricity to transmit a series of electric pulses. The longer pulses were expressed as dashes, the shorter ones dots— and each letter of the alphabet was translated into a few specific dots and/or dashes, forming the code that was transmitted by the telegrapher. For example, the letter S was expressed as three dots, or short pulses; the letter O as three dashes, or long pulses; so the message "SOS," used as a distress signal, would be three dots, three dashes, then three dots again.

Western Union was established in 1851 and, in a decade's time, had built the first transcontinental telegraph line. The first transatlantic cable offered service in 1868.

A different telegraph system using the same principles of electromagnetism was patented in Great Britain in 1837.

Samuel F. B. Morse

A line was constructed from Washington to Baltimore and Morse relayed his famous first message in 1844 from inside the Capitol Building. Within a decade, thousands of miles of wire stretched throughout the country, connecting major cities, enabling a more rapid expansion of the American West, and revolutionizing long-distance business transactions. His invention is regarded as the earliest precursor to the

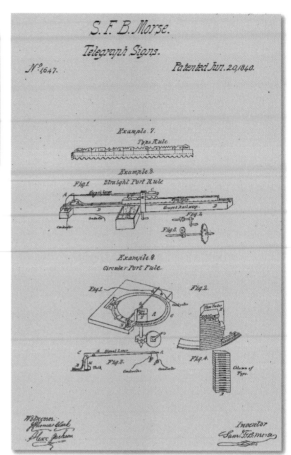

The first printing telegraph.

"information highway" or Internet. Morse's first telegraph message was ominous, and his question remains unanswered—hidden, perhaps, deeper now in the human imagination of the computer age.

How It Works

1. A circuit of electric conductors contains one or more electromagnets.

2. At the message-originating end of the circuit, an operator manipulates a signal lever that breaks and reconnects the circuit of conductors.

3. Electric pulses transmit along the circuit.

4. At the message-receiving end, a register prints out a coded language of dots and dashes derived from the short or long pulses.

5. Deciphering the code of dots and dashes reveals the messages.

"Be it known that I, the undersigned, SAMUEL F. B. MORSE, of the city, county, and State of New York, have invented a new and useful machine and system of signs for transmitting intelligence between distant points by the means of a new application and effect of electro-magnetism in producing sounds and signs, or either and also for recording permanently by the same means, and application, and effect of electromagnetism, any signs thus produced and representing intelligence, transmitted as before named between distant points; and I denominate said invention the 'American Electro Magnetic Telegraph' … "

TELEPHONE

Patent Name: Improvement in Telegraphy
Patent Number: 174,465
Patent Date: March 7, 1876
Inventor: Alexander Graham Bell of Salem, Massachusetts

What It Does

Enables people to contact one another and communicate verbally over long distances.

Background

At the age of 29, Alexander Bell was awarded this patent for an improved system of telegraphy that allowed multiple messages to be relayed simultaneously through varying harmonic vibrations. Instead of a signal lever sending interruptions over the wire, the human voice itself could send the signal. A year later, Bell established the Bell Telephone Company. His invention marked the end of the telegraph system as it was then known and the beginning of the telephone.

An early telephone operator.

Sound had been a constant interest throughout Bell's life. He was born in Scotland in 1847 to a mother who was almost completely deaf, and a father who was involved in speech pathology. Like his father, Bell became devoted to the deaf, teaching speech techniques at a school in Boston beginning in 1871. Bell was also a firm believer that science and technology could improve human life, and he possessed a natural knack for invention.

Already quite familiar with the various vibrating pitches of the human voice, he began work on what he called a "harmonic telegraph." Bell maintained that a system based on a number of transmitting instruments with different rates of vibration could simultaneously make and break an electric circuit to transmit a fluid, continuous current over a sound-carrying wire.

Bell met Thomas Watson, a man highly regarded for his abilities to improve machines who had already helped other inventors. Watson would soon help Bell improve on the finer points of his invention. On March 10, 1876, Bell spoke the first words into his machine, beckoning his assistant: "Mr. Watson, come

Bell also invented the photophone, which transmitted sound on a beam of light. The invention was a direct precursor to the fiber optics used in telecommunications today. But because Bell was unable to resolve problems associated with light source interference, the revolutionary principles of the photophone were only much later appreciated.

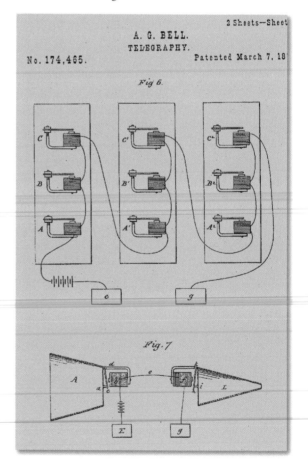

here. I want you!" In another room, Watson had received the world's first telephone call.

When Alexander Graham Bell died on August 2, 1922, telephones in the United States stopped ringing. Citizens shared a minute of silence to honor the inventor.

How It Works

1. A speaker speaks through vibrating membranes that play on the electromagnetic currents to relay an electronic facsimile of his voice over the wire.
2. Figure 7 of the patent illustration depicts an armature (c) loosely fastened to a leg (d) of an electromagnet (b).
3. The other end attaches to a stretched membrane (a).
4. A cone (A) converges sound vibrations through the membrane.
5. The vibration of the membrane enacts the motion of the armature, creating undulations upon and through the circuit (E, b, e, f, g).
6. The receiving end duplicates the vibrations so that a similar sound can be heard at another cone (L).

In the Inventor's Words

"My present invention consists in the employment of a vibratory or undulatory current of electricity in contradistinction to a merely intermittent or pulsatory current, and of a method of, and apparatus for, producing electrical undulations upon the line-wire.

TYPEWRITER

Patent Name: Improvement in Type-Writing Machines
Patent Number: 79,265
Patent Date: June 23, 1868
Inventor: C. Latham Sholes, Carlos Glidden, and Samuel W. Soule, of Milwaukee, Wisconsin

What It Does

Prints uniform characters onto paper by means of a system of push buttons that activate metal dies that impress an ink-infused ribbon onto the paper.

Background

In the 1400s, a German goldsmith named Johannes Gutenberg invented a printing press with replaceable letters that made printed materials available to the masses. It remained the standard until the twentieth century. Not until some 450 years later did a success-

A vintage Underwood typewriter.

ful personal printing press of sorts—a Type-Writing Machine—make its appearance using similar principles of moveable type. It followed hundreds of previous attempts to produce similar instruments, dating back to the early 1700s. In fact, this patent is a precursor to the one many consider to be the "first" typewriter.

Sholes was a newspaper man. He served as editor for the *Wisconsin Enquirer* in Madison, the *Milwaukee News,* and the *Milwaukee Sentinel.* In 1864, he and his friend Samuel Soule patented a page-numbering machine. Carlos Glidden, a fellow inventor, suggested that the device might be reworked into a letter-printing machine.

Glidden, Soule, and Sholes were granted a patent on June 23, 1868, for a typewriter that was an improvement of a prior invention for which they had applied for patent. A few years later Sholes and friends devel-

The "QWERTY" keyboard is named after the first six letters in the row second from the top. Its design is the result of a shrewd observation: letters placed close to each other that are commonly used together had a tendency of jamming the keys during typing. Sholes sold his patent rights for $12,000 to the rifle manufacturing company Remington & Sons in 1874, which made and marketed the machines.

American humorist Mark Twain purchased one of the early Remington model typewriters and is believed to be the first author to submit a manuscript in type to a publisher.

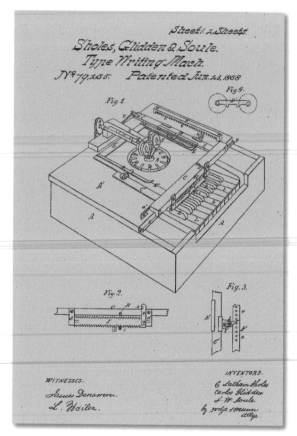

oped an even better version which would ultimately become known as the first successful commercial typewriter. The later version established the modern "QWERTY" keyboard, which remains the standard keyboard design today.

How It Works

1. When pressed, the lettered and numbered keys activate hammers with individual characters stamped on them, which in turn strike an inking ribbon against a piece of paper.

2. Each stroke engages the motion of the inking ribbon, which winds from one spool to another.

3. Also, by means of a cord and pulley system, keystrokes cause the carriage on which the paper is set to move along a set of ratchets, advancing it at each stroke until it reaches the end of a line and must be returned to the next line manually, by means of a lever.

In the Inventors' Words

"Be it known that we, C. Latham Sholes, Carlos Glidden, and Samuel W. Soule, of the city of Milwaukee, and county of Milwaukee and State of Wisconsin, have invented new and useful Improvements in Type-Writing Machines; and we do hereby declare that the following is a full, clear, and exact description of the invention, which will enable those skilled in the art to make and use the same, reference being made to the accompanying drawings, forming part of this specification."

"This invention is of improvements to an invention of a type-writing machine, an application for a patent for which we filed October 11, 1867. Its features are a better way of working the type-bars, of holding the paper on the carriage, of moving and regulating the movement of the carriage, of holding, applying, and moving the inking-ribbon, a self-adjusting platen, and a rest or cushion for the type-bars to follow."

"Thus made, the type-writer is the simplest, most perfectly adapted to its work—the writing of ordinary communications with types instead of a pen—and in every way the best of all machines yet designed for the purpose, particularly as to the cost of making the machine and the neatness and labor-saving quality of its work."

VULCANIZATION OF RUBBER

Patent Name: Improvement in India-Rubber Fabrics
Patent Number: 3,633
Patent Date: June 15, 1844
Inventor: Charles Goodyear, of New York, New York

What It Does

Provides a method of producing a durable rubber material, especially for the manufacture of auto and other tires.

Background

Sometimes a person's contributions are most celebrated after he or she is dead. Charles Goodyear's life followed this pattern. While experimenting with some raw rubber he had acquired from a shoe company, the poor hardware merchant was thrown in debtor's prison; he couldn't quite afford his investment. When he was released in 1839, he continued experimenting with rubber, adding ingredients to his recipe. He eventually produced a gummy substance in the shape of a ball, which was accidentally tossed onto a hot stove. The inadvertent process of vulcanization began. As the material melted, Goodyear identified the missing ingredient: heat. His material eventually hardened to a consistency he'd been trying to achieve from the beginning.

Vulcanized rubber met many emerging needs of the day, including a practical sole for sneakers and pneumatic tires with inner tubes for bicycles, developed by John Dunlop in 1888.

This proved to be a monumental accomplishment that has since altered the ways of the world: a pliable, sturdy rubber-based material could be molded into varying shapes and applied to any number of uses, the

most notable of which is the production of tires. Sadly, Goodyear's invention did not garner him instant fame and fortune. A company named in his honor began making synthetic tires nearly a century later, in 1937.

As Goodyear's corporate history relates it, the hot stove incident was more than a happy accident, it was a tribute to Charles Goodyear's creativity. "It held meaning for the man whose mind was prepared to draw an inference," and who had "applied himself

The first "all-weather, all-season" tire for passenger cars was manufactured in 1927 by the Goodyear Tire Company.

most perseveringly to the subject." Goodyear himself wrote: "Life should not be estimated exclusively by the standard of dollars and cents. I am not disposed to complain that I have planted and others have gathered the fruits. A man has cause for regret only when he sows and no one reaps." Charles Goodyear was a poor man when he died, but many have achieved great wealth as a result of his innovations.

How It Works

A mixture of twenty-five parts India-rubber, five parts sulphur, and seven parts white lead are heated to a temperature between 212° and 350° Fahrenheit, making a durable, flexible substance suitable for such high-impact products as car tires.

In the Inventor's Words

According to Goodyear, billions of rubber trees are now cultivated; some three million tree "milkers" harvest the crop; and the United States imports almost half of it.

"My principle improvement consists in the combining of sulphur and white lead with the India-rubber, and in the submitting of the compound thus formed to the action of heat at a regulated temperature, by which combination and exposure to heat it will be so far altered in its qualities as not to become softened by the action of the solar ray or of artificial heat at a temperature below that to which it was submitted in its preparation—say to a heat of 270° of Fahrenheit's scale—nor will it be injuriously affected by exposure to cold."

X-RAY TUBE

Patent Name: Vacuum Tube
Patent Number: 1,203,495
Patent Date: October 31, 1916
Inventor: William D. Coolidge of Schenectady, New York, assignor to General Electric

How It Works

Offers an efficient means of producing X-rays in order to capture images of the interiors of opaque bodies.

Background

X-rays are electromagnetic waves and they are very penetrative, as Wilhelm Conrad Röntgen inadvertently discovered in 1895. That fateful day, the German physicist was experimenting with a cathode-ray tube, an apparatus that displays images by means of a scanned electron beam. But Röntgen observed the effects of rays that extended beyond the range of cathode rays—rays that traveled through wood and aluminum. The rays were being generated where the electron beam hit against the inside of the tube. Because he was unfamiliar with them at the time, he called his discovery "X-radiation." His associates called it "Röntgen-radiation" or "Röntgen" rays.

Some of the benefits these X-rays offered were instantly apparent: they would provide a unique glimpse into the human body. Tissues, muscle, bones, and fat vary in their absorption of X-rays. Because they made distinct pictures of the body's bones and organs, X-rays revolutionized radiology and medicine.

For his discovery of X-rays, Wilhelm Conrad Röntgen received the first Nobel Prize for Physics in 1901. In his honor, the unit of radiation exposure equal to the quantity of ionizing radiation that will produce one electrostatic unit of electricity in one cubic centimeter of dry air at 0°C and standard atmospheric pressure is called a "röntgen."

Coolidge was awarded eighty-three patents in his life, ranging from radar systems to comfort devices such as the electric blanket. He was inducted into the Inventors Hall of Fame in 1975. One unit of electricity in one cubic centimeter of dry air at 0°C and standard atmospheric pressure is called a "röntgen."

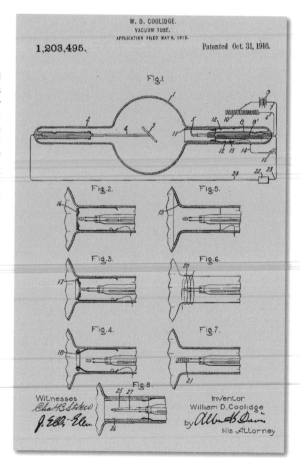

W. D. COOLIDGE.
VACUUM TUBE.
APPLICATION FILED MAY 9, 1913.

1,203,495.

Patented Oct. 31, 1916.

Fig.1.

Fig.2. Fig.5.

Fig.3. Fig.6.

Fig.4. Fig.7.

Fig.8.

Witnesses
Cha H B Stkco
J. Ell, Elen

Inventor
William D. Coolidge
by Albert H Davis
His Attorney

Subsequent efforts to develop a practical method of generating X-radiation led American electrical engineer William D. Coolidge to invent his now-famous X-ray tube. Prior to this, Coolidge, who was working for General Electric, invented ductile tungsten to be used as a refined filament now standard in electric lighting.

The widespread availability of X-rays revolutionized health care.

Previously, X-rays had been produced by bombarding an enclosed space with a steady flow of electrons which, upon colliding, underwent a sudden change in velocity and were transformed into penetrative rays. In his research with X-ray tubes, Coolidge found many problems with the existing ways of generating the radiation. In previous tubes, for example, the pressure of the gas used to affect the flow of electrons was too variable, resulting in a variable and inconsistent penetrative power of X-rays.

Coolidge decided to remove the gas used to facilitate bombardment of electrons and instead, heat a tungsten filament cathode to incandescence. He found that this would emit electrons that could be manipulated into the desired effect by electromotive force, while eliminating the variability of gas pressure.

How It Works

1. Figure 1: A glass housing (1) surrounds a plate-shaped tungsten anode (3) extended into the center by a tungsten stem (4) from a tubular end (2).
2. Lead-in wires (6, 7) are adapted to a power supply and sealed into the glass support where their energy heats cathode (5).
3. Surrounding the cathode is a hollow cylinder of tungsten (11), which acts as a static modifier to steady the focus of the cathode discharge.

In the Inventor's Words

"My invention relates to vacuum tubes and more especially to tubes operated for the purpose of producing Röntgen or X-rays. The tube which I have produced differs so radically from the tubes of the prior art used for producing Röntgen or X-rays as to amount not so much to an improvement on prior tubes as to an entirely new variety of tube differing both in its principles of operation and in its operating character."

TWO

The Little Patent That Could

AEROSOL CAN

Patent Name: Method and Means for the Atomizing
or Distribution of Liquid or Semi-Liquid Materials
Patent Number: 1,800,156
Patent Date: April 7, 1931
Inventor: Erik Rotheim, of Oslo, Norway

What It Does

Dispenses or distributes liquid or semiliquid material in "atomized" spray form.

Background

Erik Rotheim invented the forerunner of the can-and-aerosol system we now know as the aerosol spray can. He received a Norwegian patent on October 8, 1926, nearly five years before he was granted a U.S. patent for his invention. The simple practicality of this little piece of ingenuity has made it ubiquitous: Take a look in your own home and you'll surely find such necessities as spray paint, spray deodorant, hairspray, aerosol cleaning products, and lots more.

During World War II, U.S. Department of Agriculture researchers came up with a device pressurized by a liquefied gas for use as a compact bug repellent. It closely resembled today's aerosol can. Further modifications reduced or eliminated the release of fluorocarbons, which were found to damage the ozone layer.

Where there's a will there's a way, and apparently many people have a will to get high. Aerosol inhalant abuse is a serious matter with some worrisome statistics suggesting that one in five kids will have used inhalants by the time they are in eighth grade.

How It Works

1. A short pipe (7) at the bottom of a container (1) serves to fill the bottle while under pressure, being closed by compression and soldered shut when filled.

2. Near the top, a valve (4) is held in place by the action of the inner pressure on spring (8).

3. Connected to the valve is an opening (13) in the ascending pipe (3).

4. A lockable pressure knob (5) and an ejector pipe (6) serve as both a stop and a spraying orifice.

In the Inventor's Words

"When an outlet opening in the container enclosing the material and the condensed dimethyl ether is opened the material will be forced out under the pressure prevailing in the container. By suitably constructing the outlet it is possible to cause the material to pass out in the form of a solid jet or as an . . . atomized spray."

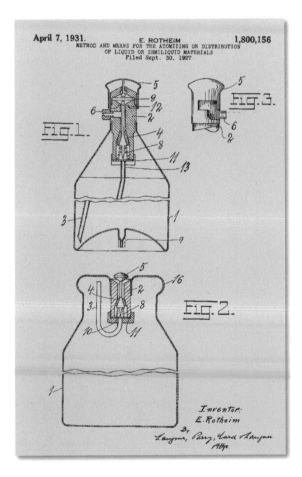

April 7, 1931.   E. ROTHEIM   1,800,156
METHOD AND MEANS FOR THE ATOMIZING OR DISTRIBUTION
OF LIQUID OR SEMILIQUID MATERIALS
Filed Sept. 30, 1927

ASPIRIN

Patent Name: Acetyl Salicylic Acid
Patent Number: 644,077
Patent Date: February 27, 1900
Inventor: Felix Hoffman, of Elberfeld, Germany

What It Does

Relieves pain associated with headache, inflammation, and fever.

Background

In the early 1800s, scientists had been able to isolate salicin, a compound found in willow plants, which was known to provide pain relief when ingested. A series of extraction methods followed, but one common complaint about ingesting salicin remained: it irritated the stomach. In 1853, a French chemist named Charles Frederic Gerhardt added sodium and chloride to alleviate the symptom but he neglected to market it and the formula became lost.

Felix Hoffman, a German chemist working for the Bayer Company, came across the work on acetyl salicylic acid several decades later. In 1899, the compound formula was marketed by Bayer under the trademark Aspirin®. It was sold in powder form until the first tablets came on the market in 1915. Bayer tried to maintain sole rights to the manufacture of the drug but was forced to relinquish its trademarks in 1919 following the Treaty of Versailles.

Felix Hoffman

Felix Hoffmann was seeking a way to ease his father's arthritis, when he invented acetyl salicylic acid.

How It Works

Aspirin belongs to a class of drugs called salicylates. It is the active ingredient in more than fifty over-the-counter medications. Aspirin inhibits an enzyme that triggers many biological processes, such as tissue inflammation.

In the Inventor's Words

If it was ever published in its time, Hoffman's patent application must have read like a slap in the face of a pharmaceutical rival: "I have now found that on heating salicylic acid with acetic anhydride a body is obtained, the properties of which are perfectly different from those of the body described by Kraut. According to my research the body obtained by means

of my new process is undoubtedly the real acetyl salicylic acid … Therefore the compound described by
Kraut cannot be the real acetyl salicylic acid, but is
another compound."

BARBED WIRE

Patent Name: Improvement in Wire-Fences
Patent Number: 157,124
Patent Date: November 24, 1874
Inventor: Joseph F. Glidden, of De Kalb, Illinois

What It Does

This wire features sharp protrusions or "barbs." Fences
made from it are an effective deterrent to whatever (or
whoever) is being kept in or out.

Background

Barbed wire is not just
used to contain livestock.
It also secures indus-
trial property, helps keep
prisoners from escaping,
safeguards temporary
stations during wartime,
and contains prisoners
of war.

Joseph Glidden

The invention of barbed wire represented a beginning
and an end in American history. It marked the end
of the free-ranging cowboy, the roving pioneer, and
the nomadic Native American tribes; it marked the
beginning of a new industry based on contained cattle-
ranching operations and it further opened the west to
settlers. Those who favored the open range vehemently
opposed the improved fencing technology. Cutting
barbed wire was a felony.

Joseph Glidden was born in Clarendon, New York,
and moved to DeKalb, Illinois, in his late twenties. By
1873, other people were already working on ways to
improve the existing simple wire fencing. It was too
easily broken by the weight of a heavy heifer. Attempts
were made to devise a pain-inducing wire that would
teach cattle to "steer" clear. The sixty-year-old Glidden
invented a form of fencing that featured barbs placed
at intervals along one wire, with another wire twisted
around the first wire, holding the barbs in place. He
received a patent for the "barbed wire" that same year
and sold half of his patent rights to Isaac Ellwood. Together, they built a factory in downtown De Kalb to
manufacture the product.

How It Works

1. Barbs are set in place along a wire.
2. This wire and a companion wire are stretched between a hook at one fence post and a thumb-piece securely fastened to another.
3. The thumb piece is designed to transversely twist the wires together, thereby securing the barbs in place.

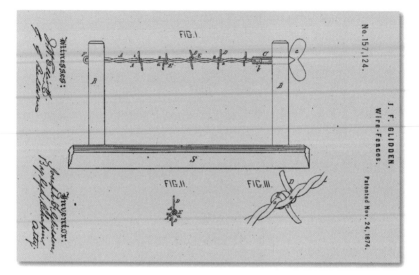

In the Inventor's Words

As native people of North America witnessed the open ranges being increasingly staked, sectioned off, and fenced in, they coined a nickname for barbed wire: the devil's rope.

"This invention has relation to means for preventing cattle from breaking through wire-fences; and it consists in combining, with the twisted fence-wires, a short transverse wire, coiled or bent at its central portion about one of the wire strands of the twist, with its free ends projecting in opposite directions, the other wire strand serving to bind the spur-wire firmly to its place, and in position, with its spur ends perpendicular to the direction of the fence-wire, lateral movement, as well as vibration, being prevented. It also consists in the construction and novel arrangement, in connection with such a twisted fence-wire, and its spur-wires, connected and arranged as above described, of a twisting-key or head-piece passing through the fence-post, carrying the ends of the fence-wires, and serving, when the spurs become loose, to tighten the twist of the wires, and thus render them rigid and firm in position."

BAR CODE

Patent Name: Classifying Apparatus and Method
Patent Number: 2,612,994
Patent Date: October 7, 1952
Inventor: Norman J. Woodland, of Ventnor, New Jersey,
and Bernard Silver, of Philadelphia, Pennsylvania

What It Does   Scans merchandise for classification and pricing
purposes.

Background   Next time you're stuck in line at the grocery store
behind someone who's paying with pennies and out-
dated coupons, count your blessings. At least you're
not waiting for the cashier to enter each item's price
manually into the cash register. That was a fact of life
when you shopped before the bar code and the bar
code scanner came into being.

The bar code system is a time-saving technology
that has significantly changed the consumer purchas-

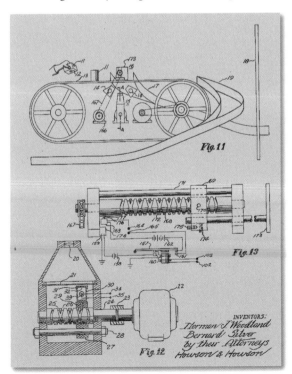

Fig. 11

Fig. 13

Fig. 12

INVENTORS:
Norman J. Woodland
Bernard Silver
by their Attorneys
Howson & Howson

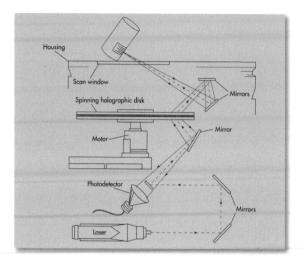

ing experience wherever it has been employed. In 1949, a local grocer asked two Drexel Institute of Technology grad students to automate product checkout. Eventually, Norman Woodland and Bernard Silver devised a system that employed a coded pattern and photosensitive technology that could read the codes. Their system is quite similar to the one used today. Their bar code had four lines. The first was a reference line and the other three specified information relating to the first line. Using more lines allowed more specific and numerous classifications to be coded. But it would be a while before industry standards were set. The bar code's first practical use was matching railroads with their correct trains.

On June 26, 1974, the first product with a bar code was scanned at a checkout counter in Troy, Ohio. It was a ten-pack of Wrigley's Juicy Fruit Chewing Gum. You can see it displayed at the Smithsonian Institution's National Museum of American History.

How It Works

1. A bar code displays relevant information through classifiable patterns of reflected and non-reflected light.
2. In Figure 11, the positions and presence of white lines against a black background represent the datum line, or bar code (12), on a product (11).
3. The highly reflective white lines move face down against a transparent conveyor (13), where they are subject to multiple sources of illumination (14).
4. The reflected emanations are immediately projected into an photosensitive optical and scanning element (15), which includes a sensing mechanism that moves back and forth across the conveyor.

Later innovations in bar code technology have employed holographic image-scanning.

"This invention relates to the art of article classification
and has particular relation to classification through the
medium of identifying patterns.

"It is an object of the invention to provide automatic apparatus for classifying things according to
photo-response to lines and/or colors which constitute
classification instructions and which have been attached to, imprinted upon or caused to represent the
things being classified."

BATTERY

Patent Name: Galvanic Battery
Patent Number: 373,064
Patent Date: November 15, 1887
Inventor: Carl Gassner, Jr., of Mentz, Germany

What It Does

Provides a portable, wireless energy source for electronically powered devices.

Background

Galvanic batteries
connected in a series to
increase the voltage.

Alessandro Volta "powered" a new field of invention
when he created the "Voltaic pile." Others before him,
including Benjamin Franklin, had tried harnessing
electricity. Volta devised the basic model: the conduc-
tion of electricity between two opposite charges. It is
the basis on which later improvements to batteries
would build. Subsequent innovations involved connecting cathodes with liquid electrodes. They were
effective, but also heavy and easily breakable.

One of the most significant modifications to the
battery was developed by Carl Gassner, who invented
the "Galvanic battery"—the first commercially successful "dry" cell battery. Gassner used a zinc container
and a porous material to absorb the electrolyte, and
sealed the cell across the top with bitumen. Gassner's
invention instantly made the battery more practical
and easier to handle. Additionally, Gassner's addition
of zinc chloride helped minimize the zinc corrosion,
extending the natural life of the battery. It is the forerunner of today's ubiquitous carbon-zinc batteries.

When a negative electrode is put in contact with a positive electrode, it causes a chemical reaction which produces electricity. In the battery, the electrodes are divided by an ion-conducting separator and housed within an ion-conducting electrolyte. The reaction is triggered when the cell is connected to an external load, conducting a current of electrons from the negative electrode back to the positive electrode.

Gassner's battery works as follows:

1. An exciting agent—the electrolyte—contains by weight the following: one part oxide of zinc; one part sal-ammoniac; three parts plaster; one part chloride of zinc; and two parts water.

2. A zinc cylinder—the container—houses an isolated cylinder of carbon manganese.

3. The space between the two cylinders is filled with the exciting agent, in liquid or semi-liquid form, which becomes relatively solidified.

In the Inventor's Words

"The oxide of zinc may be employed with great advantage as a constituent of any well-known exciting agent for the elements, with which it can be mechanically mixed and introduced into the galvanic

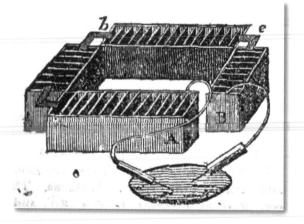

cell and act therein, as hereinafter set forth. I do not, however, claim its employment when mixed with an exciting-fluid for the electrodes and applied thereto before being introduced into the cell, so as by chemical action to transform the zinc oxide at once into a chemically-different salt—as, for instance, dilute sulphuric acid—which transforms it into zinc sulphate."

"The inner resistance of the elements will not be raised by the addition of the oxide of zinc, as the latter is a better conductor of electricity than plaster and other similar bodies, which produce only a partial and varying porosity."

Alessandro Volta was an Italian physicist known for his pioneering work in the field of electrochemistry, which led him and others to astounding discoveries in electricity. He was born into a noble family in Como, Italy, in 1745. In 1774, he was appointed to his first academic position, as professor of physics at Como High School. The very next year he came up with the electrophorous, his first significant invention. The electrophorous was made up of two plates: one was metal and covered with ebonite, and the other plate featured an insulated handle. The device accumulated the charges created by static electricity, a process that is used as the basis of electrical condensers even today. Volta's most significant contribution to the field, however, was his invention of the first crude battery in 1800.

Volta made his discovery in the most unlikely manner: He learned of the work of a scientist friend, Luigi Galvani, who, while dissecting a frog, had noticed that the dead critter's leg would twitch when touched with two different metals. Volta was intrigued by Galvani's finding, and set out to determine whether the charge was coming from the animal's tissue or from the two metal pieces used in the dissection.

Beginning in 1794, Volta began a series of experiments to try to duplicate Galvani's work. He quickly determined that the living tissue wasn't necessary for the generation of electricity; instead, he discovered that the frog's leg was merely a conductor for a charge being generated by the two metal pieces themselves. They were oppositely charged metals, so they created an electrical impulse when they were connected, in this case by the frog's leg. Volta spent the next several years trying to refine this process, and in 1800, he developed the first electric battery, which came to be known as a Voltaic pile.

Volta's battery was a far cry from the small metal batteries we are accustomed to today. The Voltaic pile consisted of round plates of copper and zinc stacked up and separated by disks of cardboard moistened with salt water. A copper wire was attached to the top and bottom of this "pile," and when the wire ends were touched, electricity flowed through the stacked disks. The units of electricity that were created were dubbed "volts" after the inventor, a term that is still in use today.

Thanks to his groundbreaking invention, Volta received considerable acclaim in his lifetime. He received the Copley medal of the Royal Society, and when the principles of the Voltaic pile became more widely known, he received many other medals and awards, including the French Legion of Honor. In 1801, he was made a count by Napoleon.

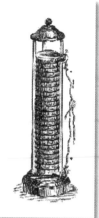

Volta's "Voltaic Pile" alternates between silver (or copper) and zinc discs, which together produce an electric current.

BOTTLE CAP

Patent Name: Bottle-Sealing Device
Patent Number: 468,226
Patent Date: February 2, 1892
Inventor: William Painter, of Baltimore, Maryland

What It Does Provides an effective seal for bottled, carbonated beverages.

Background Carbonated beverages were in popular demand by the late nineteenth century, but there were problems supplying a quality product in a bottle. Stoppers and other corking and capping implements used at the time did not reliably seal the bottles, and some had a tendency

Like stamp collecting, crown bottle cap collecting is a real passion for many people; and the Internet has brought them all a little closer. Bottle cap collectors throughout the world maintain hundreds of Web sites that showcase rare crown caps or offer trade opportunities to other collectors.

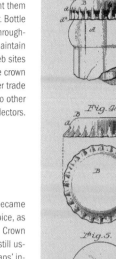

(No Model.)

W. PAINTER.
BOTTLE SEALING DEVICE.

No. 468,226.

Patented Feb. 2, 1892.

Fig. 2. *Fig. 1.* *Fig. 3.*

Fig. 4. *Fig. 8.*

Fig. 6.

Fig. 5. *B' Fig. 9.*

Fig. 10.

Fig. 7.

Attest:
Philip F. Larner.
Lowell Battle.

Inventor:
William Painter,
By   Attorney

Before plastic became the material of choice, as late as the 1960s, Crown cap makers were still using cork for the caps' inner sealing material.

Crown caps have found a niche in the art world, and are commonly used in sculpture and design. See http://www.interestingideas.com/out/cap.htm.

to either taint the flavor or flatten the carbonation. William Painter put a "stop" to these problems when he came up with his unique crown cap design in the early 1890s.

Painter's design featured not only the familiar crown shape of the cap that locked onto the lip of the bottle's neck; it also included compressible disks on the underside of the cap to provide a leakfree seal. But because his invention necessitated a change in the neck design of the bottle, it was not instantly accepted. Painter finally convinced beverage bottlers of its ingenuity and his cap went on to set the standard for bottle caps, which have changed little since. A crafty inventor, Painter received patents on more than eighty inventions, but his crown cap remained his crowning achievement.

How It Works

Two years after receiving his bottle-cap patent, Painter was issued a patent for a "bottle cap lifter," just what was needed to pop the top.

A bottle lip (a) is rounded as is the exterior packing surface (a´).

1. The recess below the lip (c´) – the locking shoulder – extends at a short, straight incline.
2. Disk (C) or (C´) is placed on top of a filled bottle. Cap (B) is then placed on top and heavy downward pressure is applied.
3. Its flared edges (d) are bent downward and inward to engage with the locking shoulder.
4. The flange design provides access to free space (e) for convenient opening with the aid of a bottle-opening device (usually a prying tool that employs leverage against the top of the cap).

In the Inventor's Words

"My present invention pertains to the sealing of bottles by the use of compressible packing – disks and metallic caps, which have flanges bent into reliable locking engagement with [ringed] locking-shoulders on the heads of bottles, while the packing disk is in each case under heavy compression and in enveloping contact with the lip of the bottle."

CALCULATOR

Patent Name: Calculating-Machine
Patent Number: 388,116
Patent Date: August 21, 1888
Inventor: William Seward Burroughs, of St. Louis, Missouri

What It Does

A high-speed device for solving simple arithmetic problems.

Background

American Arithometer factory

This invention bears almost no resemblance to the common electronic or battery-powered calculators we use today. In size, it more closely resembles a typewriter. In function, it is really an adding machine that also multiplies by making repeated additions. But this patent represents the first successful printing calculator. Two years before the issuance of this patent, Burroughs and three associates established the American Arith-

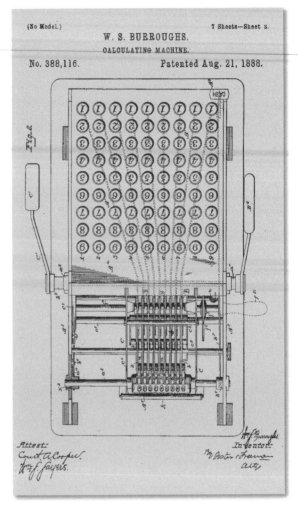

Apple Computer advertisement originally aimed at computer hobbyists.

mometer Company. They sold their first machine for a whopping $475, an almost unimaginable sum for such a device in the nineteenth century. But there was a pressing demand for what was then an extremely unusual instrument.

Previous versions of these machines were expensive, inefficient, and easily damaged. A common feature at the time was a handle that was pulled to register the numbers; varying pressures on these handles contributed to the breakage of many early adding machines. Burroughs's invention included a kind of hydraulic regulator on the handle so that no matter how sharply or gently the handle was pulled, pressure exerted on the mechanism was the same. Burroughs's calculator also printed out the calculation on a strip of paper which could be used as a record or a receipt for accounting purposes. This innovation had a major impact on the way office and business transactions were conducted.

How It Works

1. A keyboard consisting of nine columns of numbers in ascending order from 1 through 9 is arranged in nine rows.
2. The keys deploy revolving disks, which are fixed on a shaft, that contain on their peripheries figures 0 through 9.
3. An operator strikes the keys and a series of mechanized operations tallies a cumulative figure.

In the Inventor's Words

William Burroughs was posthumously inducted into the Inventor's Hall of Fame in 1987.

"My invention relates to that class of apparatus used for mechanically assisting arithmetical calculations; and my invention consists in the combination, with one or more registers, of a series of independent keys and intervening connections constructed, arranged, and operating, as fully specified hereinafter, so as to indicate upon the register the sum of any series of numbers by the proper manipulation of the keys, and also so as to print or permanently record the final result."

CELLOPHANE

Patent Names: Process for the Continuous Manufacture of Cellulose Films; Apparatus for the Continuous Manufacture of Cellulose Films
Patent Numbers: 981,368; 991,267
Patent Dates: January 10, 1911; May 2, 1911
Inventor: Edwin Brandenberger, of Vosges, France

Q: "In two thousand years, what's the greatest invention you've seen?"
A: "In two thousand years? Saran Wrap."
　　　　　　　—Carl Reiner and Mel Brooks, "The 2,000-Year-Old Man"

What It Does　Offers a method of producing a thin, strong, flexible, continuous sheet of transparent cellulose film that may be applied to a variety of practical uses.

Background　When Mel Brooks, in his legendary comedy routine "The 2,000-Year-Old-Man," jokingly paid homage to the impact of Saran Wrap, he wasn't far off the mark. But before there was Saran, there was cellophane, which opened the door to an array of innovations in plastics. At its inception, it was meant only to be waterproof.

A cellophane cover is applied to North American "Mustang" fighters in the shipping department of an Inglewood, California, plant, before they are crated and shipped to the Royal Air Force.

One story has it that, after he observed a glass of wine spilt onto a restaurant tablecloth, Swiss chemist Edwin Brandenberger decided to work on developing a stainproof fabric. Employed at the time by a French textile company, Brandenberger coated cloth with a viscose film. But while no one was interested in his product, he realized he was onto something and continued experimenting with the liquid viscose, a cellulose solution derived from plants.

On July 23, 1909, he applied for two patents: one for the process of making the diaphanous film, and another for the apparatus used to make it. In 1911, he was awarded both of these patents.

Brandenberger combined the words "cellulose" and "diaphanous" to coin the term "cellophane." The first version of the product had a greenish hue and was too expensive to be used for anything but decorative packaging for upscale merchandise. Brandenberger later assigned his patents to an organization named La

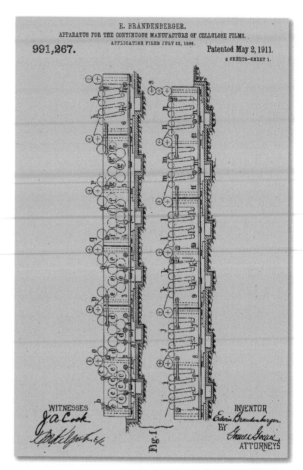

E. BRANDENBERGER.
APPARATUS FOR THE CONTINUOUS MANUFACTURE OF CELLULOSE FILMS.
APPLICATION FILED JULY 23, 1909.

991,267.

Patented May 2, 1911.
8 SHEETS—SHEET 1.

WITNESSES
INVENTOR
Edwin Brandenberger
BY
ATTORNEYS

Fig. I

> "You're the top.
> You're Mahatma Gandhi,
>
> You're the top!
> You're Napoleon brandy.
>
> You're the purple light of
> a summer night in Spain.
>
> You're the National
> Gallery. You're Garbo's
> salary,
>
> You're cellophane!"
> -Cole Porter

Cellophane Société Anonyme. This group sold patent rights to the DuPont Cellophane Company. Despite some notorious legal battles, DuPont secured a comfortable hold on the stuff, and DuPont scientists went on to perfect and market the product in a variety of forms. Dow Chemical Company soon jumped on the bandwagon, and it was this company that created Saran Wrap.

How It Works

1. A cellulose water solution, cellulose xanthum is distributed through a hopper into a solution of sulfate and ammonia. Upon contact, the first solution coagulates.

2. The coagulated film is immediately treated in a water solution of sodium chloride or sea salt to remove impurities.

3. The film is immediately directed to a third bath consisting of a mineral acid diluted with water. In its final insoluble state, the film is washed in cold water and hot water.

4. The apparatus in which the whole process can be orchestrated consists of a series of vats and rollers, as depicted in the patent illustration.

In the Inventor's Words

"This invention, relates to a process by means of which it is possible to obtain in a continuous manner films of an indefinite length, starting direct from a water solution of cellulose, more particularly from a solution of cellulose xanthate."

CHEWING GUM

Patent Name: Improved Chewing Gum
Patent Number: 98,304
Patent Date: December 28, 1869
Inventor: William F. Semple, of Mount Vernon, Ohio

What It Does

Dispenses a sugar-based "snack" while providing an active chewing experience as its mass remains in the mouth undigested.

Background

As a dentist, William Semple had the best intentions when he patented his invention: to encourage people to keep their jaws well exercised and their teeth clean. Recommended ingredients for his rubber-based "chewing gum" included chalk, charcoal, and powdered licorice root—ingredients that he believed would provide scouring properties. While Semple never aggressively marketed his invention, he did help lay the groundwork for others with names like Wrigley.

The chewing of gum was hardly an innovation in the nineteenth century. Ancient Greeks chewed *mastiche*, the resin of the mastic tree; Aztecs and Mayan Indians chewed chicle.

Meanwhile, the commander Antonio López de Santa Ana, who led the bloody charge against Fort Alamo before his army retreated, had settled on Staten Island in New York. He brought from his native land a large chunk of dried chicle, the gummy, tasteless resin of the sapodilla tree, which he enjoyed chewing. He introduced inventor Thomas Adams to his chicle;

Adams imported a large amount with the intention of making an inexpensive synthetic rubber. His experiments failed, so he decided on improving the stuff as a chewing gum.

Less than two years after Semple received his patent, in 1871, Adams began selling his flavorless chicle gumballs in a New Jersey drugstore. Adams was also the first to sell gum that could be purchased from a vending machine. Soon flavors were added and the history of gum began to take on new dimensions. The Good Dentist Semple could not have known that what he started would lead to the subsequent generations of cavity-inducing, sugar-laced gum that have become a staple to youngsters worldwide.

How It Works

Semple's gum recipe is simple:

1. Dissolve the rubber in naphtha and alcohol until it reaches a jelly-like consistency.
2. Mix in prepared chalk, powdered licorice root, or other suitable materials.
3. Enjoy, if possible.

In the Inventor's Words

"The nature of my invention consists in compounding with rubber, in any proportions, other suitable substances, so as to form not only an agreeable chewing-gum, but also, that from the scouring-properties of the same, it will subserve the purpose of a dentifrice.

"It is well known that rubber itself is too hard to be used as a chewing-gum, but in combination with non-adhesive earths may be capable of kneading into any shape under the teeth."

COLT REVOLVER

Patent Name: Improvement in Fire-Arms
Patent Number: X9430
Patent Date: February 25, 1836
Inventor: Samuel Colt, of Hartford, Connecticut

"God didn't make all men equal. Sam Colt did."
—Popular expression in the Wild West

What It Does
This easily loaded weapon contains multiple shots which can be discharged in rapid succession.

Background
Samuel Colt's invention transformed a firearm from a single-shot device to a multiple-shot device. He developed a revolving chamber; the application of caps at the end of the cylinder; a partition between the caps; a shield over the caps to protect them from moisture and smoke; and the novel principle of a connecting-rod between the hammer and the trigger. The same year he received his first U.S. patent, he established a factory in New Jersey to make the guns. But it would be six years before Colt's gun earned more than just modest attention. In 1842, his Patent Arms Manufacturing Company closed its doors.

By the beginning of the Civil War the Colt revolver was one of the world's most popular small arms.

Colt turned his attention elsewhere and invented a device that could set off explosives by remote control. But he soon returned to guns. In 1847, the doors of his business reopened when the U.S. Army entered into a contract with him to produce one thousand of his revolving-cartridge firearms for use in the Mexican War. His guns soon became the most popular during the Civil War. Colt continued honing his products and eventually produced a metallic-cartridge revolver that substantially reduced a gun's tendency to misfire. This model was revered for its accuracy. It was called the Colt .45 Caliber Peacemaker, and quickly became the gun of choice in the American West.

Using electronic ballistics technology, Australian Mike O'Dwyer invented a weapon that can fire up to a million rounds per minute. A test firing of thirty-six barrels lashed together turned fifteen wooden doors into tiny shreds in less than a quarter of a second.

How It Works

1. A hammer acts upon a fulcrum, and is drawn back by a projection.

2. A pin projection from the hammer locks the cylinder into place when it is aligned with its respective chambers.

3. Pulling the trigger from the catch of the hammer pushes a spring mechanism forward.

4. The spring activates delivery of the percussion cap into a tube where it explodes and discharges the load.

In the Inventor's Words

Samuel Colt was a true entrepreneur who understood the finer points of financing. Under the alias "Doctor Coult," he procured enough nitrous oxide to tour the country earning money demonstrating the effects of laughing gas.

"Among the many advantages in the use of these guns, independent of the number of charges they contain, are, first, the facility in loading them; second, the outward security against dampness; third, security of the lock against the smoke of the powder; fourth, the use of the partitions between the caps, which prevent fire communicating from the exploding cap to the adjoining ones; fifth, by the hammer's striking the cap at the end of the cylinder no jar is occasioned, deviating from the line of sight; sixth, the weight and location of the cylinder, which give steadiness to the hand; seventh, the great rapidity in the succession of discharges, which is effected merely by drawing back the hammer and pulling the trigger."

DRINKING STRAW

Patent Name: Artificial Straw
Patent Number: 375,962
Patent Date: January 3, 1888
Inventor: Marvin C. Stone, of Washington, District of Columbia

What It Does   Provides a means of ingesting beverages or medications through an elongated tube by the use of a sucking motion, rather than engaging directly with the container holding the liquid.

What It Does   Until the late 1880s, such natural grasses as rye or hollow reeds were used as drinking straws and as a means to administer medicine. Then straws of paper were patented for the purpose in England. The forerunner of today's drinking straw was invented by an American manufacturer of paper cigarette holders, Marvin Stone. In the days before synthetic plastic, it struck Stone that

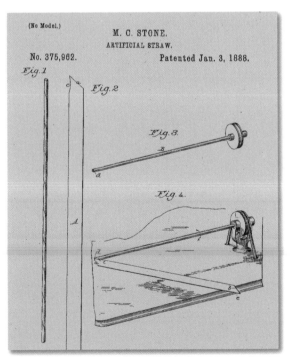

a better straw could be made by coating manila paper with paraffin and fashioning it into a tube. Eventually, a machine was devised to coil and wind the paper into tubes.

The spiral-winding technique that his company used to form the tubes may have been a more patent-worthy innovation than the straw itself. While plastic soon came to replace paper as the preferred material for drinking straws, Stone's spiral-winding method is now commonly used to make electronic components and countless other products in a variety of industries.

How It Works

1. Paper (A) is cut and formed with parallel sides and beveled ends (a, b).
2. A cylindrical spindle containing a slit (d) to receive an edge of the paper winds the paper in a spiral motion to form the tubing.
3. The tube is secured by adhesive (e) pre-applied to the opposite edge of the paper.
4. The tube is then dipped into a vat of molten paraffin to seal it and make it durable against the liquid into which it will be inserted.

In the Inventor's Words

"The aim of my invention is to provide a cheap, durable, and unobjectionable substitute for the natural straws commonly used for the administration of medicines, beverages &c.; and to this end it consists, essentially in a straw formed by winding a paper strip into tubular form and securing the final or outer edge by an adhesive material, the whole being coated with paraffin or other water-proof material."

LEVI STRAUSS BLUE JEANS

Patent Name: Improvement in Fastening Pocket-Openings
Patent Number: 139,121
Patent Date: May 20, 1873
Inventor: Jacob W. Davis, of Reno, Nevada (assignor to himself and Levi Strauss & Company, of San Francisco, California)

What It Does

Durable, comfortable pants, ideal for physical labor, include extremely sturdy seams, including those around their pockets, which are enforced with rivets.

Background

Levi Strauss & Company considers the official birthday of blue jeans as we know them today to be May 20, 1873. In reality, denim jeans were around long before then, regularly used to make durable work-wear. Levi Strauss, a native of Germany, had already established himself as a successful entrepreneur. His company manufactured denim blue jeans in San Francisco during the mining boom years of the western United States.

In 1936, the Levi Strauss Company introduced the brand-recognition concept by sewing its little red signature flag along the back pocket of its jeans. It was the first label sewn on the outside of a garment.

A Reno-based tailor from Latvia named Jacob Davis regularly ordered bolts of cloth from Levi Strauss & Company. As the story goes, one of Davis's clients returned regularly with complaints about shoddy seams, which were evidently prone to ripping. The crafty tailor solved the problem very simply with metal rivets. The small metal reinforcements could be fastened into the denim at the most critical junctures, such as the pocket corners. Davis's idea fared well with his customers, but he couldn't afford to shell out the sixty-nine bucks required to patent his idea. In need of financial backing, he partnered with Levi Strauss. The rivet concept is still used by the company, but extra-durable stitching is often substituted, partly as a concession to fashion. Far from being strictly for miners and other laborers, blue jeans have become the most popular form of casual dress in the world.

Before denim, the textile used in making sturdy pants typically came from Genoa, Italy. Weavers in France called the material "genes," which subsequently translated into "jeans" in English. Similarly, when the Italian-made textile was replaced by a softer fabric manufactured in Nîmes, France, it became known as the fabric "de Nîmes," or in American-speak, "denim."

The Levi Strauss Company has sold 2.5 billion pairs of jeans. Statistics reveal such success to be associated with customer loyalty: 97 percent of consumers own at least one pair, most own an average of seven.

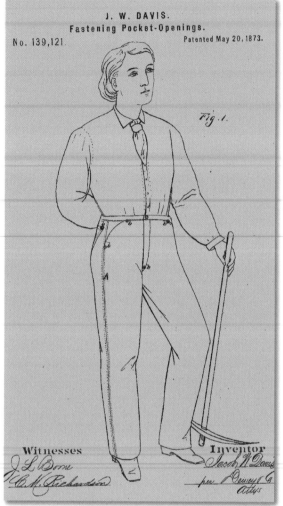

J. W. DAVIS.
Fastening Pocket-Openings.
No. 139,121.
Patented May 20, 1873.

Fig. 1.

Witnesses

Inventor
Jacob W. Davis
per Dewey & Co.
Attys

How It Works

A metal rivet is studded through a hole at the end of a seam, binding two pieces of cloth together securely. It is then "headed down" or compressed on both sides, firmly securing the seam.

In the Inventor's Words

"My invention relates to a fastening for pocket-openings, whereby the sewed seams are prevented from ripping or starting from frequent pressure or strain thereon."

OPTICAL FIBER

Patent Names: Fused Silica Optical Waveguide;
Method of Producing Optical Waveguide Fibers
Patent Numbers: 3,569,915; 3,711,262
Patent Dates: May 2, 1972; January 16, 1973
Inventors: Donald B. Keck, Peter C. Schulz, and Robert D. Maurer,
assigned by Corning Glass Works of Corning, New York

What It Does
By means of successive internal reflections, this fiber transmits information through light beams.

Background

Light travels through optical fiber more efficiently than through the atmosphere.

Optical fiber was a revolutionary invention that picked up where Alexander Bell had left off in 1880. Way ahead of his time, Bell had invented what he called a "photophone," which transmitted sound on a beam of light. The invention was a direct precursor to fiber optics used in telecommunications today. Bell, however, was unable to resolve problems associated with light source interference and so could not refine his creation. Optical fiber would prove to be the perfect mechanism for what Bell had first conceived, and for much more. A flexible, transparent fiber, usually made of glass or plastic, transmits information by successive internal reflections.

On May 11, 1970, three researchers with the Corning Glass Works filed applications for two patents: one for the optical fiber itself; the other for a method to produce it. Earlier that year, they had produced an optical fiber that possessed more than 60,000 times the data-carrying capacity of copper wire.

The three researchers were not the only ones aware of the massive potential of fiber optics. Engineers had been trying to perfect it for some time, but were unable to resolve a fundamental problem: sustaining a significant amount of light as it traveled through the fiber. The Corning trio came up with a unique waveguide innovation that did the trick. This would prove to be a monumental achievement that was first applied to long-distance telephone communications and ultimately changed multimedia telecommunications across the globe.

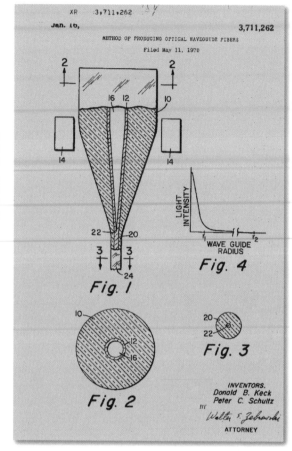

How It Works

1. Production of fiber optics begins with a rod of fused silica doped with tin oxide, titanium oxide, or several other dopant materials that increase the index of light refractivity.

2. The rod is inserted into a tube of fused silica. The temperature of the rod and tube combination is then raised until it has enough viscosity to be drawn into a wire or fiber.

3. A transmitter produces and encodes light signals. The signals travel through the core (30) of the waveguide (10) in figures 1 and 2.

4. They are refracted by the mirror-like cladding (20) made of pure fused silica or slightly-doped fused silica. An optical receiver decodes the data traveling in the light signals.

"Fused silica has excellent light transmission qualities in that absorption of light energy and intrinsic scattering of light by the material is exceptionally low. Scattering of light that does occur within fused silica is primarily caused by the presence of impurities rather than the intrinsic nature of the material itself. Furthermore, fused silica is such a hard material that an optical waveguide once formed possesses the quality of being highly resistant to damage from high temperatures, corrosive atmosphere and other severe environments."

PEZ DISPENSER

Patent Name: Pocket Article Dispensing Container
Patent Number: 2,620,061
Patent Date: December 2, 1949
Inventor: Oskar Uxa, of Vienna, Austria (assignor to Eduard Haas, of Muhlbach-Attersee, Austria)

What It Does

Offers a novel means to dispense small, square candies from a spring-loaded plastic container, appealing especially to young children for its toylike quality, and to collectors for the variety of characters depicted.

Background

In 1927, Austrian Eduard Haas III enlisted the help of a chemist to devise a cold-pressing process to make peppermints inexpensively. They designed the mints in a rectangular shape that allowed for more efficient machine-wrapping. Haas began marketing his peppermint tablets as an antismoking aid. The name PEZ comes from an abbreviation for the German word "Pfefferminz," or peppermint. Perhaps to further its appeal as an antismoking aid, dispensers mimicking the action of lighting a cigarette were patented two decades later.

PEZ dispensers come in a wide variety that delights collectors.

Haas's company thrived and when it was time to market his merchandise in North America in 1950, ad campaigns targeted kids and soon the dispensers had figure heads on them. Among the first characters to appear on PEZ dispensers were Santa Claus and Popeye. The pretense that PEZ was anything more than candy in a novelty dispenser stayed in Europe. It was the novelty that Americans couldn't resist. Today, vintage PEZ dispensers depicting characters that range from Barney Rubble to Barbie are considered hot collectibles.

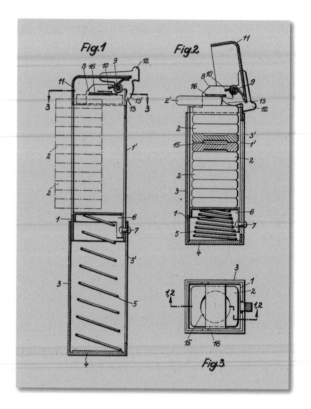

How It Works

1. An oblong plastic container with an opening at the top and a moveable base houses a stack of tablets.

2. Surrounding this container is an outer casing with a spring at its base that is set to act against the moveable base of the inner container.

3. A pin-spring hinge at the top is set to open easily upon pressure of the thumb and snap shut when released.

4. When the thumb-piece is engaged, a thrust member inside dispenses a single tablet through the opening. The removal of a tablet automatically triggers the spring to push the stack of tablets upward by the inner container's moveable base.

In the Inventor's Words

"The invention relates to pocket containers having a hinged lid and adapted for the delivery of articles, such for example as pastilles, tablets, sweetmeats, and cigarettes, which containers can be opened or closed with one hand, and present the merchandise to the user in a hygienic and unspoilt condition. The possibility of operation by one hand only is important not only for persons having only one hand but also persons who often have only one hand free (for example motor-vehicle drivers), or whose occupation causes their hands to become smeared with dirt."

TRANSISTOR

Patent Name: Semiconductor Amplifier
Patent Number: 2,502,488
Patent Date: April 4, 1950
Inventor: William Shockley, of Madison,
New Jersey, assignor to Bell Telephone

What It Does   Controls and manipulates electrical signals to maximize energy efficiency and conserve space.

Background   In conjunction with another patent (2,524,035) awarded to William Shockley's two colleagues John Bardeen and Walter Brattain for a Three-Electrode Circuit Element Utilizing Semiconductive Materials, this patent contributed to the creation of a solid-state device that could amplify electrical current—the transistor. For this great contribution, the three inventors shared a 1956 Nobel Prize. The transistor revolutionized modern electronics—replacing the vacuum tube—and became a key component in microchip and computer technology. Until the transistor's appearance, vacuum tubes were the only way to amplify electronic signals or serve as switching devices. Smaller, lighter, more durable, and more reliable than vacuum tubes, the transistor would replace the vacuum tube's function and signaled a new era in electronics technology in general, radio technology specifically, and computer technology ultimately.

But the transistor was not an overnight success. For many years, the transistor was simply not selling very well. Its price, higher than the vacuum tube, was apparently greater than its appeal. Eventually though, the trend toward miniaturization in electronics marginalized the vacuum tube as the transistor grew smaller and cheaper. This proved to be a turning point in history, when the science-fiction concept of computers became a very real feature of the workplace. Today a large portion of the world population would be unable to imagine life without computer technology. The transistor, in fact, is the technological father of the integrated circuit, which consist of microchips that link together millions of tiny transistors.

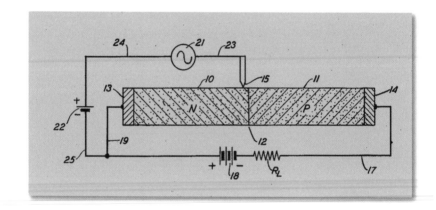

How It Works   The transistor functions dually as a tiny on/off switch and as an amplifier of electricity. Like a water faucet valve, it turns the current on and controls the rate at which it flows. A base, a collector, and an emitter make up the three basic parts of the transistor. The base acts like the control valve, opening and closing access to a larger supply of electricity provided by the collector and dispersing it through the emitter like a faucet tap.

In the illustration, a block of semiconductive material is depicted.
1. Separating the two types of electricity carriers in semiconductors (the p-type and the n-type) is a high-resistance electrical barrier (12).

2. Large area connections (13, 14) at both ends of the block are formed of solder or electroplated metal coating.

3. A conductive point contact (15) is adjacent to the barrier.

4. Current introduced by the contact causes electrons to gather and form a more efficient channel of electricity.

The materials that make up the transistor, conductors and insulators, have been essential to its success. Conductors transfer electrical current; insulators do not. The material used for the base of today's transistors—pure silicon—is considered a semiconductor. The trio of inventors knew that the semi-conductive properties of silicon, or germanium, could be exploited to create a more efficient and key component in electronics.

In the Inventor's Words

"This invention relates to means for and methods of translating or controlling electrical signals and more particularly to circuit elements utilizing semiconductors and to systems including such elements.

One general object of this invention is to provide new and improved means and methods of translating and controlling, for example, amplifying, generating, modulating, etc., electric signals.

Another general object of this invention is to enable the efficient, expeditious and economic translation or control of electrical energy.

Texas Instruments produced the first practical silicon transistors in 1954. It was used in the first fully transistorized radios to hit the market. TI also marketed the first hand-held calculator.

CHRONOLOGY OF THE COMPUTER AGE

Computer technology has progressed exponentially in the last twenty years or so, with new machines and processors coming out so fast that some machines are rendered obsolete just a few years after being introduced to the market. However, the early development of computers did not progress at such lightning speed. The first forerunners of modern-day machines were clunky, slow behemoths, and until only recently, it seemed that the giant calculating machines would be used only by trained professionals in giant laboratories.

1642: The first crude calculator—a set of wheels linked by gears—was invented by Blaise Pascal, a French mathematician, philosopher, and scientist.

Early 1670s: Gottfried Wilhelm Leibniz, a German mathematician, improved Pascal's calculator, adding components that allowed the machine to do multiplication and division. Leibniz also developed the binary system, which provided a shortcut to performing decimal-based operations.

Mid-1800s: English logician and mathematician George Boole refined the binary system, which represents data with just two numbers, 0 and 1. To this day, incredibly complex calculations are performed using software based on what is now referred to as "Boolean logic."

1801: Joseph-Marie Jacquard developed a punch-card system to automate looms. A card with a pattern created by punched-out holes was used to direct needles in the weaving process. The holes (and absence of holes) replicated the two digits—the 1s and 0s—of the binary system.

Charles Babbage

1830s: English mathematician Charles Babbage used punch-card technology to develop a mechanical computer that stored sets of punched cards for later reference in complex operations.

1890: Herman Hollerith, an American inventor and businessman, created a machine that used the punch-card system to tabulate the results of the 1890 census. His is considered the first successful computer.

1914: International Business Machines Corporation (IBM) was founded. The company is an outgrowth of Hollerith's Tabulating Machine Company.

1944: Harvard University professor Howard Aiken built the Mark 1, a digital computer controlled by electromechanical switching devices.

1945: J. Presper Eckert, Jr., and John William Mauchly, both engineers at the University of Pennsylvania, developed ENIAC, the Electronic Numerical Integrator And Computer. The giant computer took up over 15,000 square feet and was operated by 18,000 vacuum tubes.

1947: Scientists at the Bell Telephone Laboratories, led by William Shockley, invented the transistor. The transistor was smaller, lighter, more durable, and more reliable than vacuum tubes, and its creation signaled a new era in computer technology. The transistor's main function is to rapidly turn on and off the flow of electricity, and to amplify the flow when needed.

1951: Eckert and Mauchly joined forces with John von Neumann, an early pioneer in the field of computer logics, to develop EDVAC, the Electronic Discrete Variable Automatic Computer, an influential precursor to later computers. Later that year, Eckert and Mauchly also completed the UNIVAC, the UNIVersal Automatic Computer, which became the first commercially viable computer.

1958: American engineer Seymour Cray designed the first computer powered entirely by transistors.

1959: IBM unveiled its first transistorized computer.

1959: The earliest integrated circuits were introduced, otherwise known as computer chips (or silicon chips). The miniaturization of computer technology began.

Late 1960s: Businesses begin to rely more heavily on computers and link them into networks.

Late 1960s: The U.S. government's Department of Defense developed a way to link all of their computers together as a preventive measure in case of attack or natural disaster. The network was called ARPAnet. Eventually, businesses, universities, and other institutions linked their own systems into this nationwide network, creating what is known today as the Internet.

Early 1970s: Computers became smaller as developments in chip technology made it possible for more and more information to fit into smaller and smaller spaces.

January 1975: The Altair 8800 computer—a build-it-yourself kit based on the Intel 8080 processor—was sold via mail order by Model Instrumentation Telemetry Systems, a company in Albuquerque, New Mexico. *Popular Electronics* magazine made it their cover story. The name comes from the television show *Star Trek* and was suggested by the daughter of MITS's owner, Ed Roberts.

1977: Steven P. Jobs and Stephen G. Wozniak were intrigued by the Altair, but couldn't afford one so they built their own. They went on to develop the Apple II personal computer. The relatively inexpensive machine was easy to use and practical for regular people who didn't have specialized computer training. Small business, families, students, and schools began to purchase computers.

1981: IBM introduced its version of the personal computer. It soon surpassed the Apple II in popularity.

1984: Apple bit back with the Macintosh, a simple-to-use desktop computer embraced by the public for its easy-to-use graphics and design programs.

1991: Englishman Tim Berners-Lee developed Hyper Text Mark-up Language (HTML), then used the new code to launch the World Wide Web. The Web introduced graphics, video, sound, and other features to the Internet for the first time. Berners-Lee also devised the universal address system used on the Web, otherwise known as the Uniform Resource Locator, or URL, which gives each Web site its unique address.

1993: Marc Andreessen, a student at the University of Illinois, began development of Mosaic, the first Web browser, designed to easily display HTML documents. Mosaic formed the basis of the popular Netscape browser Andreessen went on to develop in Mountain View, California.

Late 1990s: The World Wide Web becomes an immensely popular common medium used by people around the world. E-mail, Web pages, online retail, and chat rooms all become part of the popular vernacular.

2003 and beyond: Computers facilitate the operation of nearly every technological wonder of our time. With computer power doubling nearly every two years, some experts predict that computer intelligence may soon surpass the human ability to think and reason. The idea is both exciting and frightening—but only time will tell how far computer technology will go.

VELCRO

Patent Name: Velvet Type Fabric and Method of Producing Same
Patent Number: 2,717,437
Patent Date: September 13, 1955
Inventor: George de Mestral, of Prangins, Vaud, Switzerland

What It Does

This self-fastening material is made of tiny hooks on one side and loops on the other, which can be separated and reattached many times.

Background

Style 511
As Shown

An unconventional use of Velcro.

In 1948, the Swiss inventor George de Mestral returned from a walk in the woods with his dog to find himself and the dog covered in cockleburs. Intrigued by what made them stick, he looked at the burrs under a microscope and saw the natural loop-grabbing hooks protruding from the core. He decided to invent a fabric based on the adherent characteristics of the burr. He teamed up with a French weaver and, several years later, perfected what he called Velcro, a combination of the words "velour" and "crochet."

One of the most important little inventions in textile history, Velcro has found countless practical uses. In addition to replacing zippers and buttons on clothing, shoes, and wallets, Velcro has endless more pragmatic purposes. Its appeal lies in its ability to stick together instantly and be torn open quickly—accompanied by a satisfying sound. It is even used to keep equipment in place aboard space shuttles.

How It Works

Today, "Velcro" is both a trademarked product and the name of the company that makes it.

1. A foundation material is constituted by a weft strand (1) and a warp strand (2).
2. Warp threads (2, 3) are woven to form a raised pile (9, 10).
3. Threads (9) are bent down at the ends, forming a hook (4).
4. When the hook side meets the pile side of the velcro, the hooks stick into the pile, forming a tight and tidy seal.

"My invention has for its object a velvet fabric including a foundation structure constituted by a weft and a warp incorporating threads that are cut at a predetermined length so as to form a raised pile. My novel fabric distinguishes from the other similar fabrics by the fact that the raised pile is made of artificial material, while at least part of the threads in said pile is provided near its end with material-engaging means, as required for adhering to a similar fabric or for scouring purposes."

Velcro is not used strictly for practical purposes. It is, for example, an important component in such human-performed stunts as the favorite run-jump-and-stick-trick. This involves a suit of Velcro, and a Velcro-covered wall.

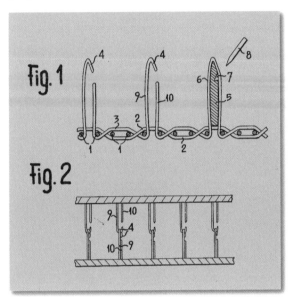

ZIPPER

Patent Name: Clasp Locker or Unlocker for Shoes
Patent Number: 504,038
Patent Date: August 29, 1893
Inventor: Whitcomb L. Judson, of Chicago, Illinois

What It Does A strip made of metal or synthetic material features "teeth" that interlock when drawn together, providing a simple and useful fastener for articles of clothing, handbags, or other fabric items.

A Japanese company called Yoshida Kogyo Kabushikikaisha became the largest supplier of zippers. Hence, the "YKK" engraved on zippers of all kinds. The world's largest zipper factory is in Macon, Georgia, and produces seven million zippers a day.

In 1851, Elias Howe received a patent for an earlier zipperlike clothing enclosure. Had he marketed his device, it may have secured his place in history as the creator of *two* noteworthy inventions: Howe had already invented the sewing machine.

(No Model.)

W. L. JUDSON.
CLASP LOCKER OR UNLOCKER FOR SHOES.

No. 504,038.

Patented Aug. 29, 1893.

Fig. 1. *Fig. 2.* *Fig. 6.* *Fig. 8.* *Fig. 9.* *Fig. 10.* *Fig. 7.* *Fig. 5.* *Fig. 3.* *Fig. 4.*

Background

In 1917, electrical engineer Gideon Sundbeck received a patent for a "separable fastener," which was a considerable improvement upon Whitcomb's. It included an increased number of interlocking parts per inch, set in two rows of teeth.

Like many great inventions, the zipper came about as a response to a physical challenge. Whitcomb Judson was inspired to invent the mechanism as a result of his large physique. Simply put, Judson had difficulty tying his own shoes.

At the 1893 Chicago World's Fair, Judson showcased his clasp locker mechanism on a shoe. Apparently, not many people were impressed. It was considered clunky and displeasing to the eye. The interlocking clasps were far apart and the ring to operate the mechanism was rather large. As a replacement for the normal buckle or string system, the clasp locker was just too conspicuous. Nevertheless, the mechanism sewn into the unsightly footwear displayed some of the defining principles of the modern zipper.

1. Hooked clasps with interlocking parts engage only when at an angle to the line of strain.
2. Overlapping or under-reaching projections at the forward ends of the clasps prevent disengagement of the hooks.
3. A moveable guide that includes cam-ways for the clasps allows for the engagement or disengagement of the clasps in a single motion.

In the Inventor's Words

"The invention was especially designed for use as a shoe fastener; but is capable of general application wherever clasps consisting of interlocking parts may be applied, as for example, to mail-bags, belts, and the closing of seams uniting flexible bodies."

ZIPPO LIGHTER

Patent Name: Pocket Lighter
Patent Number: 2,032,695
Patent Date: March 3, 1936
Inventors: George Gimera and George G. Blaisdell, of Bradford, Pennsylvania, assignors to Zippo Manufacturing Company

What It Does

This portable, pocket-sized, reusable device provides flame, particularly for igniting smoking materials such as cigarettes or cigars.

Background

The first Zippo lighter was introduced in 1932 and cost the consumer $1.95. This was a hefty price to pay in the days of the Depression. But George Blaisdell, later known as Mr. Zippo, was determined to manufacture and market a quality product. One of his brilliant marketing tactics was to offer a lifetime warranty. In a time before the majority considered smoking a bad habit, a lifetime warranty on a sharp-looking, wind-proof lighter was an attractive feature.

Today, the Zippo lighter says two things of its owner: you take your smoking seriously; and you take your style seriously. When a stranger at the train depot asks for a light, nothing is cooler than responding with the nonchalant flip of a Zippo lid and an instanta-

Nearly 400 million Zippo lighters have been produced since 1933.

March 3, 1936. G. GIMERA ET AL 2,032,695
POCKET LIGHTER
Filed May 17, 1934

Fig.1

Fig.3

Fig.4

Fig.2

Fig.5

WITNESSES
E. J. Maloney.
V. A. Peckham.

INVENTORS
George Gimera
George G. Blaisdell
By Brown, Critchlow & Flick
ATTORNEYS

Since the company's beginnings, Zippo has made more than 325 million windproof lighters.

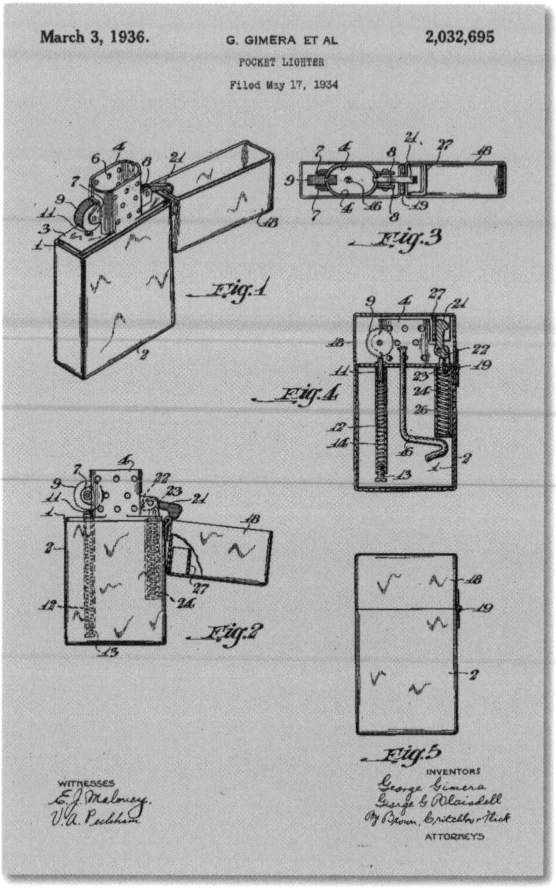

neous snap of flame. With its chic and compact design, the refillable lighter also becomes a personal keepsake.

How It Works

War correspondent Ernie Pyle wrote about the high demand for the Zippo in World War II, calling it "the most coveted thing in the Army."

1. A casing houses two disposed hollow members (1, 2) that fit snugly against one another.

2. A top plate (3) encloses the inner housing.

3. At the top center of the plate, an elliptical wind screen with draft openings (6) emerges.

4. On one side, a toothed wheel (9) is supported to engage an elongated flint (11).

5. A small hinge (19) at the side of the top plate opposite the toothed wheel attaches to the cover (18) of the outer casing.

6. A casing houses two disposed hollow members (1, 2) that fit snugly against one another.

7. A top plate (3) encloses the inner housing.

At the top center of the plate, an elliptical wind screen with draft openings (6) emerges.

8. On one side, a toothed wheel (9) is supported to engage an elongated flint (11).

9. A small hinge (19) at the side of the top plate opposite the toothed wheel attaches to the cover (18) of the outer casing.

10. When the lid is opened, a downward flick of the thumb on the wheel creates a spark against the flint. The spark is ignited into flame by a wick that extends from within the lighter's interior and the lighter fluid tank.

11. Light up and take a drag. Don't inhale the first drag: you don't need that butane.

In the Inventor's Words

"Pocket lighters having covers hinged on their upper ends must have means for holding them closed if they are to be satisfactory, and it is also desirable that they have means for preventing the covers from closing prematurely and extinguishing the flame after they are opened. In lighters known heretofore these means take the form of exposed latches and interior springs and levers which take up space in the lighters and reduce their storage capacity for inflammable lighter fluid. Furthermore, the exposed latches are apt to catch in and wear the clothing, to be accidentally actuated in the pocket, and to accumulate dirt. Such lighters are complex and frequently get out of order, while they must be constructed in a cheap manner in order to be saleable at a low price."

THREE

The More Humane Patent

ADHESIVE BANDAGE

Patent Number: 2,823,672
Patent Date: February 18, 1958
Inventors: Peter Schladermundt, of Bronxville, New York, and William H. Dennerlein, of Beechhurst, New York; assigned by Johnson & Johnson

What It Does

Band-Aid bandage strips are used around the world to treat minor burns and cuts. The sterile gauze pad protects the wound and absorbs blood while promoting healing.

Background

The Band-Aid story actually goes back thirty-eight years before this patent was issued, to a young newly-wed couple living in Brunswick, New Jersey, in 1920. Earle Dickson worked for Johnson & Johnson, a company that specialized in the manufacture of surgical dressings. Josephine was a housewife, prone to accidents in the kitchen. Earle found himself regularly tending to her cuts and burns when he was struck with the idea of placing pieces of gauze onto strips of tape in anticipation of his wife's next mishap.

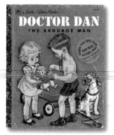

This Litttle Golden Book came with bandages.

Earle soon convinced his employers that the idea of these ready-made bandages might have commercial appeal; Johnson & Johnson began making the first adhesive bandages, cautiously, in 1921. The handmade strips were 3 inches wide, 18 inches long, and did not enjoy overnight success—but eventually, the idea stuck. In 1924, the company installed machines for mass producing the new product, and the trade name Band-Aid was adopted. Earle was made a vice president of the company, presumably allowing him to take his wife out to dinner a little more often.

In 1951, the first Band-Aid brand plastic strips were introduced. This patent is a close cousin to what most of us would recognize as the Band-Aid we use now. Specifically, this patent solves some of the previous problems of the product (e.g., the peel-away strips had too much contact with the adhesive around the sterile pad, causing potential for inadvertent contact between the pad and a person's nonsterilized finger).

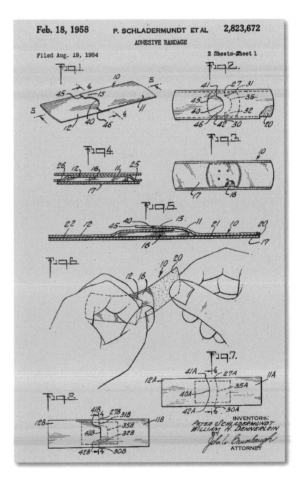

Feb. 18, 1958   P. SCHLADERMUNDT ET AL   2,823,672
ADHESIVE BANDAGE
Filed Aug. 19, 1954   2 Sheets-Sheet 1

How It Works   1. An adhesive bandage has a top facing strip and a bottom facing strip.

2. The bottom strip overlaps absorbent sterilized gauze.

3. A backing strip contains adhesive on its top.

4. The gauze pad divides the backing strip into two exposed areas on opposite endwise portions of the strip, which can be quickly peeled apart at the time of application.

In the Inventors' Words

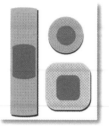

"Adhesive bandages, the term being used herein to designate not only the elongated strips, but also the round 'spots' and the relatively square 'patches,' are formed of a backing sheet which may be cloth or plastic coated on one side with an adhesive and having a dressing pad secured to the adhesive, leaving exposed areas of adhesive on both ends of the dressing. The endwise adhesive tabs generally constitute the major area of exposed adhesive. However, even in the elongated strips appreciable areas of exposed adhesive often appear at the sides of the dressing, although the latter are generally smaller, relatively speaking, than the areas of adhesive exposed at the ends of the dressing. The sidewise exposed adhesive areas are proportionately larger in the case of the round 'spot' and the square 'patch' dressing, in which case, they are usually as large as the endwise adhesive areas. Accordingly, the present invention is directed to the 'spot' and 'patch' dressings as well as the elongated strip adhesive bandages."

AIR BAG

Patent Name: Air bag restraint system with venting means
Patent Number: 5,071,161
Patent Date: December 10, 1991
Inventors: Geoffrey L. Mahon, of Ridgewood, New Jersey, and Allen Breed, of Boonton Township, New Jersey

What It Does Prevents injuries and saves lives during automobile accidents by inflating upon impact and creating a buffer between driver or passenger and the car's interior, windshield, etc.

Background In 1968, Allen Breed invented an automotive sensor and safety system that employed the world's first electromechanical automotive air bag system. By 1973,

GM was testing some of their first air bags. The automotive industry was already well aware of mounting consumer concerns over driver and passenger safety, but public demand was not so great as to demand that air bags become a regular feature in automobiles; and the industry was in no hurry to add more expense to its overhead.

When, eventually, the air bag did become a standard feature, it saved many lives. But injuries arising from the air bag itself were commonly reported as well. Tests and reports by accident victims revealed that, in many cases, the pressure of the air bag upon deployment was too high. The bag, meant to cushion the impact of a crash, was creating its own forceful impact. Using the resources of his automotive technology company, Breed modified his invention by adding a ventilation system to his air bag that would buffer the velocity of the impact while still protecting the passenger. Warnings were added against subjecting small children or the frail to the impact of the air bag.

How It Works

1. Crash sensors are located in several areas on the front of a vehicle.
2. The sensors connect to an air bag assembly through a wiring harness.
3. During a collision, the sensors send an electric spark to an inflator canister.
4. Within the space of a second from the time the sensors go off, nitrogen gas inflates the air bag at a rate of between 130 and 200 mph.
5. Vents in Breed's improved air bag allow slight deflation of the gas at a preselected rate to eliminate injuries from secondary collisions of a passenger with the bag.

In the Inventors' Words

"In an ideal situation, a constant restraint load is applied at the instant the impact begins and is such that the occupant utilizes all the available distance within the occupant compartment. Such a force would be the minimum required to dissipate the occupant's energy.

"In the worst case, no restraint force is applied until the vehicle is fully deformed and has come to rest, and the occupant has completely traversed the compartment. At this point the occupant will then impact the steering wheel, windshield and/or instrument panel. Depending upon the compliance of these components (which, in general are not very compliant and also tend to exert localized loading), the occupant will sustain his energy change by way of a very high force exerted over the relatively short distance these components will yield.

"What can realistically be achieved lies somewhere in between these two extremes."

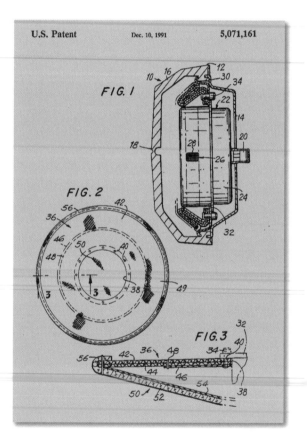

U.S. Patent   Dec. 10, 1991   5,071,161

FIG. 1

FIG. 2

FIG. 3

AIR CONDITIONING

Patent Name: Apparatus for Treating Air
Patent Number: 808,897
Patent Date: January 2, 1906
Inventor: Willis H. Carrier, of Buffalo, New York

What It Does   Decreases the temperature of indoor air by drawing in air from outdoors, cooling it, and then circulating it throughout a room by means of fans and blowers.

Background   Willis H. Carrier's air-conditioning epiphany came while he waited for a train one uncomfortably humid evening. A moment of clarity revealed to him the correlation between temperature, humidity, and dew point. He soon created the first air conditioner. Carrier's invention was first put to use in a Brooklyn printing plant, where fluctuations in temperatures warped the paper. As a result, a frustrated printer couldn't get the colors aligned.

A central office air condtioning unit.

After this success, the young engineer was awarded his own subsidiary company to make the device for businesses and warehouses, where large machinery required cooling or the products they made required a way to keep them from melting. Countless product lines suffered as a result of extreme and varying conditions in the air. The textile industry became a major supporter since certain fabrics were prone to deterioration in uncontrolled temperatures. Air conditioning was especially useful in cotton mills where a lack of moisture caused static electricity in the fibers, which were difficult to weave.

In 1915, Carrier and a small group of others established Carrier Engineering Company, which promised to deliver systems that could regulate temperature and humidity levels as desired by their industrial customers. It was not until 1924 that Carrier's ideas were applied to cooling air for personal comfort rather than industrial needs. Many people felt their first blast of AC in department stores and movie theaters, which had installed them to keep their customers coming in the summertime.

The Depression slowed the success of the air conditioner, but nor for long. The postwar housing sprawl that popularized a suburban lifestyle also boosted the demand for air conditioning. Today, for most people living in developed nations, it is hard to imagine life without the luxury of a regulated indoor atmosphere.

The company that Carrier started in 1915 is now an affiliate of United Technologies and conducts business in more than 170 countries.

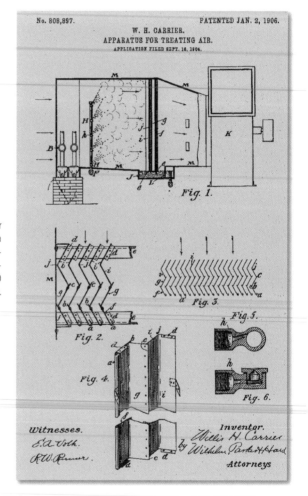

No. 808,897.

PATENTED JAN. 2, 1906.

W. H. CARRIER.
APPARATUS FOR TREATING AIR.
APPLICATION FILED SEPT. 16, 1904.

Fig. 1.

Fig. 2.

Fig. 3.

Fig. 4.

Fig. 5.

Fig. 6.

Witnesses.

Inventor.
Willis H. Carrier
by Wilhelm Parker & Hard
Attorneys

1. A fan or propelling device (K) sends air current through the air trunk (M).

2. Attached to a supply pipe (F), a spraying device (H) is situated near the air trunk intake.

3. Sprayed air travels through a separator made of baffle plates to form a series of deflections to remove moisture and impurities in the treated air.

4. Flanges on the plates (b, c) keep the moisture trapped against the plates while the air continually makes contact with it in its sinuous path through the separator.

5. A basin (J) and filter (L) collect the moisture at the bottom. Inside the casing, pipe coils (B) provide a heating or cooling mechanism to control the air temperature.

In the Inventor's Words

Willis Carrier

"This invention relates to apparatus for treating air previous to its use for ventilating and heating buildings or for other commercial purposes—such as drying, refrigerating, &c.—and more particularly to air-purifying apparatus of that kind in which a liquid or solution in a finely-divided condition or atomized spray is introduced into a current of air to be treated, which is then caused to pass through a separator consisting of baffle-plates which intercept and separate from the air the particles of liquid, together with the solid impurities contained therein."

ARTIFICIAL LEG

Patent Name: Artificial Leg
Patent Number: 37,282
Patent Date: January 6, 1863
Inventors: T. F. Engelbrecht, of New York, New York; R. Boeklin, of Brooklyn, New York; W. Stahlen, of Brooklyn, New York

What It Does

Provides a viable, comfortable, and aesthetically acceptable substitute for a lost leg.

Background

As represented in the model photo, this Civil War prosthetic may look like part of an early design for the flamboyant *Star Wars* robot, C3PO. But this improvement on artificial limbs was a significant ancestor to modern prosthetics, and it proved extremely useful during the Civil War, a bloody time in American history that claimed the lives of more than 600,000 and made amputees of some 100,000 more. Up until the time that artificial limbs of this sort were available, amputees had to rely on ill-fitting, often painful, and usually unsightly "peg legs."

How It Works

1. The thigh-piece (A) can be laced to fit closely onto the stump of the natural leg while simultaneously allowing ventilation to the healthy tissue.
2. The thigh-piece connects to the lower leg (B) and upper portion of the ankle-joint (C), to which the foot (D) is connected.
3. The knee joint is replaced here by means of a joint-pin (E); and a spring (G) is applied to straighten the knee-joint, mimicking the movement of a biological limb.
4. A ball -(j) and-socket ankle joint is employed with a spring to allow the foot independent movement while maintaining its proper position in relation to the leg.

There are many complex and ingenious principles employed in this patent, both in its cost-effective production and in the product itself. For example, the design even accounted for subtle variations in the arch of the foot. The inventors had come up with an adjust-

Original patent model.

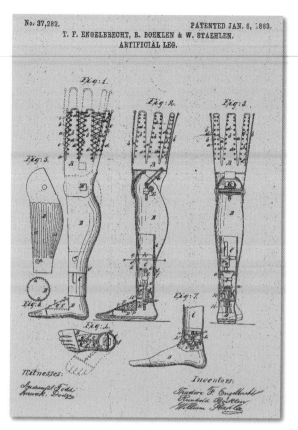

No. 37,282.   PATENTED JAN. 6, 1863.
T. F. ENGELBRECHT, R. BOEKLEN & W. STAEHLEN.
ARTIFICIAL LEG.

In 1946, the development at the University of California at Berkeley of a suction sock for above-knee prosthetics was a milestone in the attachment of lower limbs.

able artificial limb that could be mass-produced at low cost while offering a strong and lightweight substitute for a missing limb. They proposed using sheet metal wrought into shape for the thigh and lower leg, connecting the edges by soldering. Such thinking advanced the field of modern prosthetic science.

In the Inventors' Words

"This invention consists in certain provisions for the adjustment of the parts of an artificial limb in such manner as may be desirable to adapt it to the length of the natural limb and conformation of the foot of the intending wearer, by which means the necessity of making a limb to suit each particular case is to a great extent obviated, and in consequence the cost is considerably reduced; also, in a certain novel construction of the ankle-joint, by which the movements permitted to the said joint are made to approximate nearer to the natural joint than it does in the artificial legs heretofore constructed, such construction being also appli-

Manufacturing artificial legs.

cable to the wrist and other joints of an artificial arm and hand; and it further consists in the construction of an artificial limb as are required to be rigid, of corrugated plate metal, whereby the limb can be made of sufficient strength with less weight and more durable when made of the materials heretofore employed."

BRA

Patent Name: Brassière
Patent Number: 1,115,674
Patent Date: November 3, 1914
Inventor: Mary P. Jacob, of Mamaroneck, New York

What It Does   Confines, shapes, and supports the bust comfortably, with a minimum of fabric—especially in the back—to facilitate its use under form-fitting, revealing or low-backed garments.

Background   In 1913, Mary Phelps Jacob said good-bye, once and for all, to the uncomfortable corsets of the day. These undergarments were made with steel rods and whale bones. They did not conform naturally to a woman's body, nor were they well adapted to shifting fashion

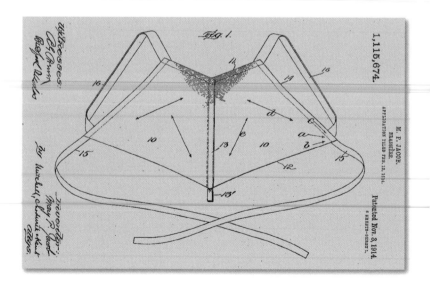

trends. While she was getting ready for an evening on the town, the inventor gazed at her own image and saw that the beauty of her brand-new dress was badly compromised by the poking framework of the underlying corset. So instead of wearing the corset, she took a couple of silk handkerchiefs and some ribbon and made herself a substitute. Though it did not consist of the double-cupping design that would ultimately become the prototype, Jacob's resourceful innovation would soon provide the basis for the first widely accepted contemporary brassière.

How It Works

1. In figure 1, two pieces of angle-cut cloth (10) are joined together by a seam (13) that extends vertically across the middle.
2. The top edges of the cloths are trimmed shorter than the bottom edges.
3. Elastic shoulder straps (16) are sewn onto the top and bottom outer edges of the two cloths.
4. Two ties or strings are sewn into bindings from the top outer edges of each cloth to the bottom, with a significant length (15) remaining.
5. The angle at which the two ribbons are sewn allows for a low fastening behind the wearer's back.

In the Inventor's Words

"The garments, in order to serve their purpose with low corsets, require to be snug fitting in order to shape the figure properly above the waist to confine the bust and conceal the corset top. Garments hitherto proposed for this purpose have required lacing or other fastening of parts across the back of the wearer or have been otherwise arranged so high as to interfere with the wearing of evening gowns cut low in the back. The necessity of a snug fit has also made it necessary that the brassiere be made with a special regard to the size and shape of the wearer in order to obtain a proper fit."

It was the Russian immigrant Ida Rosenthal who founded Maidenform in 1928 and categorized bras into cup sizes.

I dreamed I was

WANTED
in my Maidenform bra

'FRAME-UP' new bra with 3-way support
Embroidered panels frame, outline and separate the cups. Extra-firm supports at the sides give you extra uplift. Stretch band at the bottom keeps the bra snug and securely in place. It's a 'Frame-up'—in A, B, C cups.

$1.59

A 1950s advertisement

IBOT

Patent Name: System and Method for Stair
Climbing in a Cluster-Wheel Vehicle
Patent Number: 6,311,794
Patent Date: November 6, 2001
Inventors: John B. Morell, of Manchester, New Hampshire; John
M. Kerwin, of Weston, Massachusetts; Dean L. Kamen, of Bedford,
New Hampshire; Robert R. Ambrogi, of Manchester, New Hampshire;
Robert J. Duggan, of Strafford, New Hampshire; Richard K.
Heinzman, of Francestown, New Hampshire; Brian R. Key, of Pelham,
New Hampshire—assigned by Deka Products Limited Partnership.

What It Does

This sophisticated version of a wheelchair is intricately balanced and powered so that it can go where no wheelchair has gone before—up and down stairs, over curbs—virtually anywhere a mobile person can go. Its easy "joystick" operation also allows for eye-to-eye contact with the world.

Background

Dean Kamen's name has become inextricably linked with his much-publicized invention, the Segway scooter. That product may not have caught on quite as readily as his supporters believed it would, but Kamen will be remembered for many other significant and humane patents as well. Before inventing the Segway, Kamen was immersed in improving medical technologies and has a long list of credits and patents to his name. After establishing the DEKA Research & Development Corporation in Manchester, NH Kamen and a team of engineers developed the IBOT, which employs many of the same technological principles which later was used for the Segway.

Kamen with the Ibot

With two pairs of midsize wheels on a swivel and a complex set of computer-controlled gyroscopes that continually self-balance, the IBOT is an extremely sophisticated machine that offers paraplegics an incredible alternative to traditional wheelchairs, and nearly as much freedom to move around as an able-bodied person enjoys. Controlled by a joystick, it can climb curbs and go up and down stairs, and can be raised to give chair-bound people an eye-to-eye view with someone standing tall. Kamen introduced his product through Independence Technology, a subsidiary company of Johnson & Johnson, which now sells the machine.

How It Works

A significant departure from all previous wheelchair innovations, the IBOT carries a considerable price tag, selling at around $29,000 per unit. Of course, given the freedom of movement the machine provides, this cost may seem inconsequential—to those lucky enough to afford it.

The IBOT employs wheel and cluster control laws that relate to the location of the center of gravity. A computer constantly monitors the angle at which the vehicle travels. The front and rear angles of the wheels are constantly updated, so that during stair climbing, for example, the rider can vary the position of the vehicle's center of gravity with very little effort. As a safety feature, a brake pitch control monitors the cluster and wheel motor temperature during stair climbing. If temperatures exceed a preset value, the brake pitch controller turns off the motor amplifiers and controls the motion by modulating the cluster brakes. The controller places the vehicle in a configuration where all four wheels are on the stairs thus placing the vehicle in a statically stable configuration with respect

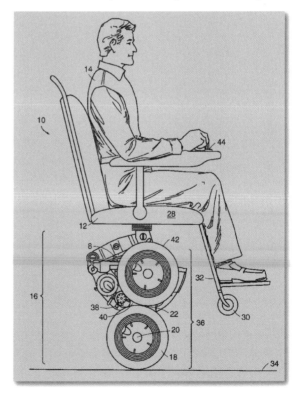

to gravity and preventing the rider from continuing in stair mode. If the motors cool sufficiently, IBOT will resume stair mode in order to allow the rider to complete the ascent or descent.

1. The operator's seat (12) is attached to a ground-contacting module (16), which contains a power source, drive amplifiers, drive motors, and a controller.
2. These components drive a cluster (36) of wheels (18).
3. The wheels are mounted on a cluster arm (40) and each wheel is capable of being driven by a controller (8).
4. The cluster arm rotates on an axis (22), the rotation of the arm governed by the controller.
5. The controller comprises a processor in communication with a memory storage device, which executes a control program.
6. A number of sensors (38) detects the precise state of the vehicle and receives commands from the rider's control stick (44).
7. These sensors can detect the pitch, roll, and yaw of the vehicle (10) as well as such variables as the angular position and/or rate of rotation of the wheels and cluster arm.
8. The independent control of the clusters and wheels allows the vehicle to operate in several modes thereby allowing the rider or the controller to switch between modes in response to the local terrain.

In the Inventors' Words

"The device has a plurality of wheels rotatable about axes that are fixed with respect to a cluster arm, where the cluster arm itself is rotated about an axis so that wheels rest on successive stairs. The wheels and cluster arms are controlled according to separate control laws by a controller. Whether the device ascends or descends the stairs is governed by the pitch of the device relative to specified front and rear angles. Shifting the center of gravity of the combination of device and payload governs the direction of motion of the device."

Dean Kamen, an inventor, engineer, and physicist, has a passion for science and technology, and high hopes for what they can accomplish. Born in 1951 on Long Island, he attended Worcester Polytechnic Institute. While still in school, Kamen made his first significant contribution to medical technology and to society. He invented a wearable device that administered small, precise amounts of medication over time. His innovation allowed patients who relied on consistent or periodic injections of medication to move about freely without being tied down to a machine. The idea changed the lives of thousands who had previously been immobile.

On the strength of this invention, Kamen went on to found DEKA Research & Development Corporation, a laboratory that incubated many more ideas related to medical research and technology. One of the lab's more notable inventions is a portable dialysis machine that allows people with kidney disease to receive dialysis at home.

Dean Kamen with President Bill Clinton

Kamen's inventive energy often focused on ways to make life easier for people around the world. The Segway is a two-wheeled, upright vehicle with an electric motor that can go 12 miles an hour. Kamen hopes the Segway will become an alternative to more polluting forms of transportation, like automobiles. He envisions a world in which people in cities will use Segways instead of cars to get around, causing levels of pollution and congestion to go down—though so far, they have been slow to catch on.

Kamen is a visionary, and he puts a lot of faith in young people. He created an interactive learning center called Science Enrichment Encounters in Manchester, New Hampshire. To help inspire and encourage young scientists, he also founded the nonprofit organization FIRST (For Inspiration and Recognition of Science and Technology) in

1989. FIRST conducts a national robot-building competition and offers over $1 million in scholarships to budding scientists and engineers. Kamen firmly believes that technology can be harnessed to solve many of the world's problems like famine, pollution, the destruction of the ozone, and the contamination of the environment. He is currently working on a device that purifies drinking water, which will help eliminate water-borne diseases. He hopes to distribute the final product to poor families in third world countries. With his dedication to kids and the environment, and his infectious passion for science, Kamen has been the recipient of numerous awards and honors. He continues to dream up inventions designed to improve the quality of life on Earth.

KEVLAR

Patent Name: Wholly Aromatic Carbocyclic Polycarbonamide Fiber Having Orientation of Less than About 45°
Patent Number: 3,819,587
Patent Date: June 25, 1974
Inventor: Stephanie Kwolek, of Wilmington, Delaware

What It Does

Kevlar's unique internal fiber structure and exceptionally high tensile properties make it ideal for manufacturing the resilient and lightweight body armor used by law enforcement officials and the military.

Background

Stephanie Kwolek

Stephanie Kwolek was a young chemist employed in 1950 by DuPont's Pioneering Research Laboratory in Wilmington, Delaware. Her experiments during the 1960s resulted in the development of synthetic fibers much stronger than any created before. In fact, the material she invented is five times stronger than its weight in steel. DuPont instantly found a commercial use for Kwolek's new crystalline polymers: Kevlar "bulletproof" vests. Hundreds of lives have been saved annually as a result of this invention.

Kevlar has received a lot of publicity recently in light of world affairs, and the need for bulletproof fabric in the fight against terrorism. Kwolek, now retired, is proud to have contributed an invention that has helped preserve human life. She remains devoted to improving the world through science. "I can't see how anybody can live in this modern world without an understanding of science," she says, "especially because it touches every aspect of our lives."

How It Works

1. Polyparaphenylene terephthalamide is produced through a chemical process called polymerization.
2. The crystalline liquid with polymers is extruded from a spinneret, a metal plate studded with tiny holes.
3. The fibers that emerge through the holes are cooled in water to allow them to harden.
4. The fibers are subsequently twisted together into yarn that can be woven into a durable, synthetic cloth.

In the Inventor's Words

There are many other significant uses for Kevlar, including strengthening suspension bridge cables, making bomb-resistant aircraft baggage containers, manufacturing safer and stronger car tires, and even making durable sails for ships.

"This invention relates to novel, optically anisotropic dopes [absorbent or adsorbent material] consisting essentially of carbocyclic aromatic polyamides in suitable liquid media. These dopes, and related isotropic dopes, are used to prepare useful fibers, films, fibrids, and coatings. In particular, fibers of unique internal structure and exceptionally high tensile properties are provided."

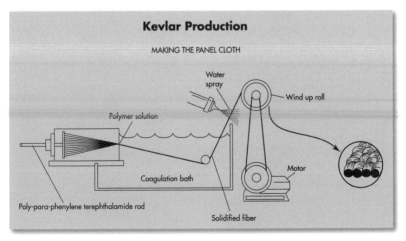

Kevlar Production

MAKING THE PANEL CLOTH

Water spray

Wind up roll

Polymer solution

Motor

Coagulation bath

Poly-para-phenylene terephthalamide rod

Solidified fiber

LASER EYE SURGERY

Patent Name: Far Ultraviolet Surgical and Dental Procedures
Patent Number: 4,784,135
Patent Date: November 15, 1988
Inventors: Samuel E. Blum, of White Plains, New York; Rangaswamy Srinivasan, of Ossining, New York; and James J. Wynne, of Mt. Kisco, New York

What It Does Gets rid of your glasses, for one thing. The ultraviolet surgical and dental procedures described in this patent broke new ground with the use of a laser that photo-etches organic biological matter without using heat.

Background When he was a doctoral student at New York's Columbia University in the '50s, Gordon Gould first coined the term "laser"—an acronym for "light amplification by stimulated emission of radiation." The word was only partly original: he'd been studying under the pro-

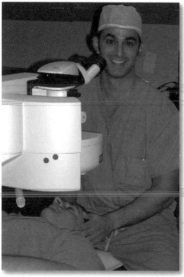

LASIK eye surgeon.

fessor who invented "masar" (microwave amplification by stimulated emission of radiation). His new term would soon overshadow his professor's. Gould believed that light-wave amplification had a lot more potential, and power, than microwave amplification. It turned out that he was correct.

By the 1980s, laser technology was being used in a variety of fields, including eye surgery. In this field, lasers were only used to create therapeutic scar tissue, until three IBM scientists—the awardees of this patent—began using a short-pulse ultraviolet laser called the excimer laser, which—unlike previous lasers—did not generate substantial heat. It did not burn tissue during surgery and it was also much more precise. It opened a whole range of surgery based on a new ability to improve, rather than simply to repair.

LASIK, which stands for laser in-situ keratomileu-sis, is one such example. The excimer laser is used to permanently change the shape of the cornea to correct flawed vision. The surgery has already helped millions of people all over the world to put down their glasses and contact lenses for good.. The impact LASIK surgery has had on the traditional corrective lens industry would make an interesting study.

Another application of the invention is in the field of dental medicine, in which the far-ultraviolet radiation of wavelengths can be used to remove decay from teeth without damaging the surrounding enamel.

How It Works

1. The patent illustration shows a source (10) of ultraviolet radiation of wavelengths less than 200 nm, such as the excimer laser.
2. A casing (12) is used to contain the laser beam (14).
3. The casing includes a shutter (16), which can either block the radiation beam or allow it to pass.
4. A reflecting mirror (18) is used to change the direction of the radiation beam.
5. A lens (20) can be used to focus the radiation beam onto a selected spot of the organic material (22). An aperture (24) provides more precise alignment of the radiation before it strikes the lens.
6. The instrument is attached to a moveable arm with articulated joints to so that a surgeon can move the beam around in precise manner.

In the Inventors' Words

"This invention relates to surgical and dental procedures using far ultraviolet radiation of wavelengths less than 200 nm, and more particularly to a method and apparatus for selectively removing organic material without heating and damage to surrounding organic material."

MAGNETIC RESONANCE IMAGING

Patent Name: Apparatus and Method for Detecting Cancer in Tissue
Patent Number: 3,789,832
Patent Date: February 5, 1974
Inventor: Raymond V. Damadian, of Forest Hills, New York

What It Does Analyzes tissue for presence and degree of cancerous malignancy, serious muscle strain, and other important medical diagnostic information.

Background Magnetic Resonance Imaging (MRI) is a means by which soft tissue, including musculature, is photographed to detect certain structural pathologies and anomalies. The technique relies on the magnetic properties of hydrogen atoms, which are present in most organic molecules and water in the human body. When the hydrogen atoms are immersed in a powerful magnetic field and bombarded with radio waves, they emit radio signals that provide information about their local environment. MRI technology provides doctors with an entirely new and profoundly significant view inside the human body.

MRIs are used in the U.S. for more than one million scans each year. Many people experience the anxiety of claustrophobia at the thought of going into the dreaded MRI tube. A more legitimate fear is of what might fly toward the tube when the magnetic field is turned on. Though relatively safe as a diagnostic machine, the powerful magnetic force of the MRI apparatus has caused accidents, including a tragic one in which an oxygen tank flew forcefully into the magnetic field, killing a six-year-old boy.

The 2003 Nobel Prize in Science was awarded to Paul C. Lauterbur, of the University of Illinois, and Sir Peter Mansfield, of the University of Nottingham in England, who are acknowledged as MRI inventors, despite Damadian's previous involvement. The two men made significant improvements upon Damadian's original idea—including the employment of a magnet gradient principle—considered distinct innovations in their own right. Embittered, Damadian placed news ads denouncing the Nobel committee's decision as an oversight. But the Nobel committee has never reversed a decision, and since deliberations are confidential, Damadian will never know if he was ever seriously considered.

One thing is certain: His work was crucial to the development of the technology used today. In the early 1970s, Damadian reported in the *Journal of Science* that the radio signals emitted by cancerous tissue were detectable as markedly different from those of healthy tissue. He envisioned the day when MRI, then called nuclear magnetic resonance, would be able to detect these signals. He drew blueprints for such a machine and received several patents, including this one. Damadian says that he essentially laid the groundwork for the discovery—and he is not alone. In 1989, Damadian was inducted into the U.S. National Inventor's Hall of Fame, which cites him as the creator of "the magnetic resonance imaging (MRI) scanner, which has revolutionized the field of diagnostic medicine."

How It Works

The first MRI exam performed on a human being took place on July 3, 1977.

The MRI scanner has three basic functions: magnetizing, applying radio waves, and using imaging technology to draw pictures. First, a part of the body in question, or the entire human body, is subjected to powerful magnetic fields. Inside the human body, millions of tiny atoms—which happen to have magnetic properties—line up with the magnetic field created by the machine. Secondly, radio waves are applied into the field to tip the atoms out of line with the field, simultaneously imposing energy into them. When the radio waves are removed from the field, the atoms reposition themselves using the energy imparted by the radio waves. In doing so, they emit subtle radio signals that are picked up by a highly receptive antenna. Deformities can be detected by fluctuations in the radio signals. Lastly, a computer registers the radio signals and produces a black and white image.

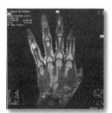

An MRI of the hand.

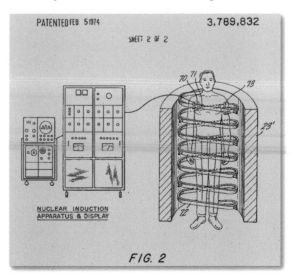

1. In figure 2, an electromagnet (23) is displayed in cross-section.
2. A transmitter probe is fixed on a track (72) that can slide along the housing in a helical pattern for access to various body parts.
3. The transmitter probe includes a beam focusing mechanism (71) that concentrates magnetic energy from a radio frequency generator.
4. A similarly arranged receiver probe (73) detects and transmits the magnetic energy.
5. The induction apparatus registers images based on the view obtained by the magnetic probes.

In the Inventor's Words "This invention relates to an apparatus and method of detecting cancer in tissue and, more particularly, to the use of nuclear magnetic resonance techniques to detect alterations in the organization and structure of selected nuclei in the tissue which alterations are believed to be caused by cancer."

NICOTINE PATCH

Patent Name: Transcutaneous Application of Nicotine
Patent Number: 4,597,961
Patent Date: July 1, 1986
Inventors: Frank Etscorn

"He lit the cigarette and smoked it down to the filter in one breath. He silently thanked the cigarette company for being thoughtful enough about his health to include a filter to protect him. So he lit up another."
—Steve Martin, "The Smokers"

What It Does Designed to ease the withdrawal process when quitting smoking, the patch provides a steady, controlled release of nicotine through the skin, without the tars and toxins of actual cigarettes.

Background After receiving a doctoral degree in experimental psychology from George Peabody College in Nashville in the mid-1970s, Frank Etscorn moved to the Southwest to teach at the New Mexico Institute of Mining and

Studies reveal that nicotine itself actually has medicinal properties known to have positive effects in the treatment of Attention Deficit Disorder, Parkinson's disease, Alzheimer's disease, Tourette's syndrome, ulcerative colitis, and schizophrenia.

Technology. While here, he used a favorite ingredient of local fare—the spicy capsicum or chili pepper—to illustrate a theory suggesting that not only are humans able to turn painful experiences into pleasurable ones, they can develop cravings for the natural chemicals that the body releases in the process. "We need a fix of red or green chilé with a side order of endorphins," he reported in 1990 in the *Albuquerque Journal.*

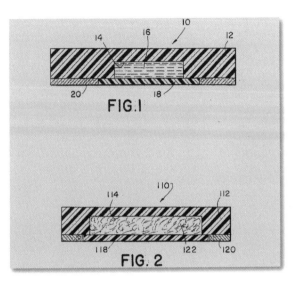

The Outlaw of Marlboro Country: Both Nicotrol and Nicoderm are the commercial products of co-inventor Jed Rose, who is often credited as the inventor of the patch. Rose is a chief researcher at Duke University, located in Durham, North Carolina, near many major cigarette companies.

Others have expanded on this pain-to-pleasure principle to uncover the nature of addiction and reveal how our species is perhaps the only one that willfully ingests what are otherwise intolerable substances, such as nicotine. Dr. Etscorn had the serendipitous experience of ingesting the drug—not willingly, by accident!—when he spilled some liquid nicotine in his lab in 1979. He could feel the substance being absorbed through his skin, or "transcutaneously," and this led to his invention of the nicotine patch.

Sipping a cig? Among the most recent smoking cessation aids currently under development, a one-time-use straw is being designed to deliver tiny beads of nicotine while its user sips a beverage of his choosing. Presumably, there is a psychological correlation to the physical similarities between straws and cigarettes.

While science has shed great light on the hazards of smoking, the hard-to-kick habit is still a leading cause of disease and death throughout the world. In 1988, the U.S. Surgeon General's Office released a report decisively stating that tobacco products containing nicotine are, indeed, addictive. By the mid-1990s, smoking cessation was an industry in itself. Gums, mints, and other nicotine substitutes or supplements, including the patch, were all aggressively marketed. The patch was considered by many experts—and consumers—the most effective. There are several major brands on the market including Nicoderm, Prostep, and Habitrol, the one Etscorn's invention established. Nixing its prescription requirements in 1996, the United States Food and Drug Administration approved over-the-counter sale of nicotine patches.

How It Works

A measured dosage of nicotine is administered through a tiny reservoir fixed into an absorbent adhesive pad that is applied to the skin. Dosages are lowered over time, gradually diminishing nicotine dependency until it is gone. In the illustration, a nicotine-infused bandage (10) has a diameter of about 1.5 centimeters and a thickness of about 2.0 centimeters. A nicotine impermeable backing (12) made of impervious plastic or rubber material contains a cavity (14) along one surface to contain the liquid nicotine (16). Sealing the nicotine film (18), again is a suitable flexible material inert to the nicotine. Adhesive (20) is provided around the nicotine permeable membrane.

In the Inventor's Words

"Nicotine appears to be the most pharmacologically active substance in tobacco smoke, yet it appears to be not as significant from a health standpoint as the tars and carbon monoxide. However, nicotine is very important from another standpoint, i.e., it is the reinforcing substance in tobacco which maintains the addiction. In this respect, a theme commonly heard among workers in the field of smoking research is, "People would be disinclined to smoke cigarettes if an alternate route of nicotine delivery could be devised."

"Several such attempts have been made to administer nicotine in alternate ways, but with varying and generally ineffective results. For example, nicotine-containing pills have been studied; however, effective blood levels of nicotine are not achieved. Drugs absorbed in the stomach pass through the liver first where, in this case, 80–90 percent of nicotine deactivation occurs. Similar findings have been demonstrated with nicotine chewing gum although it has been sufficiently successful to warrant its marketing."

PARACHUTE

Patent Name: Life-Saving Device for Aeroplane
Patent Number: 1,332,143
Patent Date: February 24, 1920
Inventor: Karl O. K. Osterday, of Cincinnati, Ohio

What It Does

Provides a safe, low-impact means of descending from great heights (as in exiting an airplane while in flight) by deployment of a fabric canopy that uses wind resistance to slow the pace of descent and reduce the force of impact.

Background

The idea of floating from a great height to land safely on terra firma has long occupied humans. Leonardo da Vinci sketched a version of a parachute some five  hundred years ago, with the idea of using it to rescue people from burning buildings. Since then, many have made and employed primitive versions of parachutes, including a Frenchman who dropped a dog to safety in a 'chute following a hot-air balloon mishap. World War I expedited the development of improved parachutes that could be easily contained and quickly employed in emergency situations involving warplanes. Karl Osterday was by no means the father of the parachute—in fact, his invention was not a favorite among aviators, who preferred detached packs rather than those worn on the back. Nonetheless, some of the principles Osterday developed for this patent led directly to subsequent improvements on the parachute, including multiple canopies and a mechanism to allow the user to free himself of the parachute upon landing. It is a device that has saved countless lives in both war and peace.

How It Works

1. Belts and straps are fastened around an aviator's waist, torso, and thighs, forming a secure harness.
2. The straps connect at the back to a canvas pack containing two of three parachute canopies.

K. O. K. OSTERDAY, DEC'D.
P. OSTERDAY, ADMINISTRATOR.
LIFE SAVING DEVICE FOR AEROPLANES.
APPLICATION FILED OCT. 16, 1918.

1,332,143.

Patented Feb. 24, 1920.

Parachuting or "skydiving," in a variety of forms, is a popular recreation for adrenaline junkies across the world. On April 18, 2000, 588 parachutists from five different countries were dropped from seven separate aircraft to set a world record for the largest mass parachute jump.

3. A third canopy is contained in a smaller pack mounted on the airplane itself. A slip-knot is fastened to a ring on top of the aviator's helmet, and is attached to the aviator's wrist by means of a cord.

4. If a pilot finds loses control of the airplane, he pulls the cord (29) so as to open the receptacle (27), and then quickly jumps out of the plane.

5. Canopies (21, 22) are dragged out of the receptacle by cords (24) and immediately fill with air.

6. These canopies also engage the cords to activate the other canopy.

7. As the pilot begins his landing, he can disengage himself from the entire parachute by pulling on a cord (18).

In the Inventor's Words

"One object of the present invention is to provide a novel and improved parachute device by means of which the aviator can release himself from his wrecked machine, while in the air, and make a safe landing.

"Another object is to provide a novel and improved device of this character wherein the aviator can release himself from the parachute when within a desired and safe distance from the ground, so that he will be unencumbered upon landing."

PROZAC

Patent Name: Aryloxyphenylpropylamines
Patent Number: 4,314,081
Patent Date: February 2, 1982
Inventor: Bryan B. Molloy, of Indianapolis, Indiana, and Klaus K. Schmiegel, of Indianapolis, Indiana; assigned by Eli Lilly and Company

What It Does This psychotropic medication, taken in pill form, has proved a godsend for many sufferers of chronic depression by inhibiting reuptake of serotonin in the brain, and thereby altering mood and state of mind without causing hallucinations or other harsh side effects.

Background In his memoir *Darkness Visible*, novelist William Styron recounts his own personal struggle with a condition that nearly claimed his life. Indeed, depression is a very serious medical condition, too often with fatal consequences. While skeptics persist in disparaging

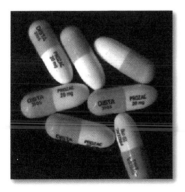

what they do not know, neuroscience, psychological research, and gifted insightful writers such as Styron are helping move the condition from beneath the social shadows of stigma to shed light on the faulty brain chemistry plagues countless numbers of people. The fact that there may be a large population who will jump at the chance to feel better (whether they really need to or not) should not be a negative reflection on those who genuinely benefit from the help that antidepressants offer; nor should it discourage those who feel they need such help. Enough stigmatism remains.

One revolutionary breakthrough in the treatment of depression came with the discovery of selective serotonin reuptake inhibitors (SSRIs). Serotonin is a chemical messenger secreted by one nerve cell and picked up by another. It is believed that many people suffer the symptoms of depression because serotonin is absorbed too quickly. Prozac became the first in a growing class of commercial antidepressant medications known as SSRIs, which are believed to increase serotonin levels in the brain. "Aryloxyphenylpropylamines" include the active ingredient in Prozac, the compound known as fluoxetine hydrochloride.

In the past, antidepressant medications addressed dopamine and norephinephrine, other compounds in the monoamines group. Along with these, serotonin plays significant roles in mood regulation, eating behavior, aggression, and pain. The success of antidepressants has helped to prove that depression is caused by an imbalance of neurotransmitters in the brain.

How It Works

Prozac was first introduced in the U.S. in 1988. It has since been approved in more than ninety countries and used by more than 40 million people.

In the limbic area of the brain, serotonin is blocked from re-absorption on the synapses between cells. However, brain chemistry is still an extremely mysterious and largely unknown frontier. Little is known about exactly how in balancing the neurotransmitter levels it achieves results for patients. Experiment and research cautiously continue in the study and development of the field. Doctors will often gradually increase doses to judge the success of the drug and to minimize side effects.

In the Inventors' Words

Depression does not discriminate; it afflicts all kinds of people from all walks of life. American Psychological Association statistics show that approximately one out of every five Americans can expect to get some form of depression, and that every year more than one in twenty have a depressive disorder.

"The tricyclic antidepressant drugs presently being marketed inhibit the uptake of monoamines by brain neurons, most of them being more effective in inhibiting the uptake of norepinephrine. Many of the compounds of this invention behave similarly in that they block norepinephrine uptake more effectively than they do serotonin uptake. Exceptions are p-trifluoromethyl derivatives, the dimethylamino, monomethylamino and unsubstituted amine derivatives being far more effective in inhibiting serotonin uptake than in inhibiting norepinephrine uptake. Thus, although the compounds of this invention clearly have potential as anti-depressant compounds, it is apparent that

N-methyl 3-(p-trifluoromethylphenyloxy)-3-phenyl-propylamine and its tertiary and primary amine analogs will have a different type of anti-depressant action from the presently marketed drugs."

READING SYSTEM FOR THE BLIND

Patent Name: Reading System
Patent Number: 6,199,042
Patent Date: March 6, 2001
Inventor: Raymond C. Kurzweil, of Newton, Massachusetts

What It Does

Translates written words into audible speech.

Background

Invented in 1976, the Kurzweil Reading Machine is the world's first computer to transform text into computer-spoken words. It scans print and uses technology that recognizes letter patterns to "read aloud" the material on the page, thus providing the blind and visually impaired access to virtually anything in print. The Kurzweil Reading Machine is considered the most significant advance for the blind since Braille was developed almost 150 years prior.

Stevie Wonder nominated Kurzweil for the nation's most prestigious award in invention and innovation, the $500,000 Lemelson-MIT Prize, which Kurzweil received. The inventor and the famous musician-composer forged a friendship that inspired another Kurzweil invention: a system of computer-based instruments that could realistically reproduce the sounds of acoustic instruments, such as the grand piano.

The system is an example of the enormous benefits that new technologies (computer chips and optical recognition software) can yield. And it is only one of the creations on this patentee's impressive resume. In addition to the first print-to-speech reading machine for the blind, Kurzweil helped to develop the first omni-font optical character recognition method, the first CCD flatbed scanner, a text-to-speech synthesizer, and the first commercially marketed large-vocabulary speech recognition system.

How It Works

1. The reading system includes a computer, a mass-storage device, and software that allows the machine to "read" an optically scanned image of a document.
2. The software converts image files into text files, and relays positional information associating the text with its image representation.

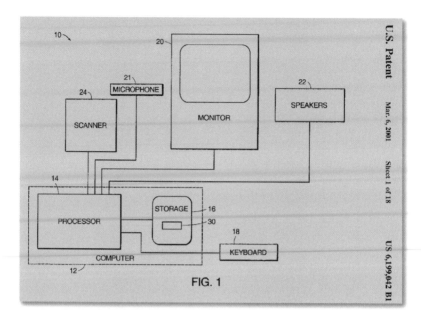

FIG. 1

Kurzweil was inducted in the Inventors Hall of Fame in 2002.

3. In synchronization with a highlighting indicator applied to a displayed representation of a document, the computer selects from a series of prerecorded voice samples.

4. While highlighting is applied to the words on the monitor, the reading machine plays back the stored, recorded voice samples.

In the Inventor's Words

Raymond Kurzweil

"In one aspect of the invention, a computer program product residing on a computer readable medium includes instructions for causing a computer to display a representation of a document on a computer monitor. The product also causes the computer to read the displayed representation of the document by using a recorded human voice. Optionally the recorded human voice is synchronized with highlighting applied to the displayed representation. The computer program product uses information associated with a text file to synchronize the recorded human voice and the highlighting to the displayed representation of the document."

RESPIRATOR

Patent Name: Respirator
Patent Number: 3,191,596
Patent Date: June 29, 1965
Inventor: Forrest M. Bird, of Palm Springs, California
and Henry L. Pohndorf, of El Cerrito, California

What It Does This pulmonary aid assists people with breathing problems by duplicating the functions of the human respiratory system.

Background Before the modern respirator, there was the iron lung, a behemoth medical device that looked vaguely like a prehistoric whole-body MRI scanner. A patient would lie inside of the machine to receive treatment for respiratory paralysis, an often fatal symptom of poliomyelitis. (Rare today, polio was rampant before Jonas Salk developed a vaccine to combat it in 1952.) Encased in the chamber, the patient's entire body would be subjected to the pressure of pump-activated bellows that would "breathe" air into the chamber, activating subatmospheric pressure on the patient's chest and, in turn, enabling the patient to breathe.

These machines went the way of the dinosaur after Forrest Bird's innovative spirit swept through. Bird invented the first mass-produced medical artificial ventilation unit: the Bird Mark 7. Forrest Bird held several patents, three of which pertained to the revolutionary nature of his respirator: this one; patent number 3,068,856 for a "Fluid Control Device"; and patent number 3,842,828 for a "Pediatric Ventilator." The

Iron lung, Carrie Tingley Hospital for Crippled Children.

last one, which significantly reduced respiration-related infant mortality, was respectfully nicknamed, and trademarked, the Baby Bird.

Another pun on his name could be made in connection with his early career as a jet and helicopter pilot during World War II. Airplanes were then ascending to elevations never attained before, and Bird became interested in the technology used to help

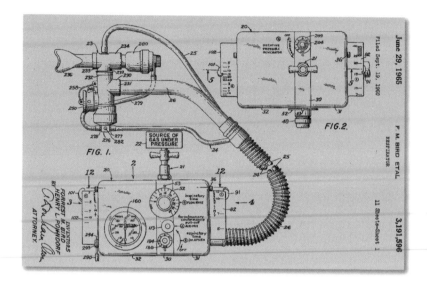

June 29, 1965

Filed Sept. 19, 1960

F. M. BIRD ETAL

RESPIRATOR

11 Sheets-Sheet 1

3,191,596

FIG. 1.

FIG. 2.

SOURCE OF GAS UNDER PRESSURE

INVENTORS
FORREST M. BIRD
BY HENRY L. POHNDORF
ATTORNEY

deliver oxygen to pilots at high elevation. Bird then steered his career toward medicine and cardiopulmonary physiology, establishing a company in 1954 and going on to develop some of the most successful respirators known.

How It Works   Bird's respirator is a complex machine. More than 350 parts are numbered in eighteen separate patent drawings. Yet, while it served a number of highly sophisticated functions, it weighed only about six pounds and had the dimensions of a small shoebox, making it an attractive alternative to the gargantuan iron lung.

1. During its operative cycle, the device works in two phases: an inhalation phase and an exhalation phase.
2. A controller attaches by an inlet to a source of pressurized gas and an outlet to connect to a patient's airway.
3. A main valve controls the flow of gas from inlet to outlet.
4. It remains open during the inhalation phase and closes during exhalation by means of a pressure-sensing mechanism for a predetermined duration.

In the Inventor's Words

"Mechanical ventilators impose their design limitations on the patient, some respirators not even offering simple adjustments for combating bronchiolar collapse, and impeding or encouraging venous return. An important object of this invention is to provide a respirator that always provides peak pulmonary conformance, so that the physician, freed from having to devote attention to this problem, can devote his attention to considering respiratory patterns for good tidal exchange, even distribution of alveolar air, facilitation of bronchial drainage, and other adjustments."

SEATBELT

Patent Name: Safety Belt
Patent Number: 312,085
Patent Date: February 10, 1885
Inventor: Edward J. Claghorn, of New York, NY

What It Does

Provides a safety restraint mechanism to keep an individual in any movable seat. It was later adapted for the seats of powered vehicles such as automobiles and airplanes, to counteract the forward momentum caused upon impact or in case of turbulence.

Background

The first safety belt was devised more than 110 years ago, primarily for workmen who needed to ascend or descend in tall buildings. This was before the rise of the automobile, and in retrospect, the invention seems visionary. The use of safety belts in cars and airplanes has saved countless lives and is required by law in many places throughout the world—but its original purpose was more general. According to the patent application, the safety belt was "intended for the use of tourists, sailors, painters, farmers, firemen, telegraph-men, and others."

How It Works

Restrains motion and locks one securely into a stationary position.

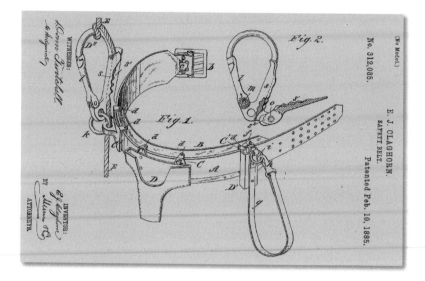

Fig. 2.

Fig. 1.

WITNESSES:

INVENTOR:
BY
ATTORNEYS.

A Swedish safety engineer named Nils Bohlin invented what is recognized now as the modern safety belt—the three-point seatbelt. Not surprisingly, he was working for Volvo, which prides itself on its strong reputation for safety. Volvo was the first automobile company to make safety belts a regular feature of its cars in 1959.

1. A leather outer belt (A) of sufficient length includes at one end a buckle (b) and fastens around a person's waist through typical belt-hole perforations.

2. The outer belt is attached to a wider, protective inner belt (B), which extends nearly throughout the length of the outer belt where it is firmly sewn or otherwise attached with sufficient open space between the belts

3. Bands (C, C1, and C2) are fastened to the outer belts and include rollers (d) on their backs to allow them sliding movement along the outer belt.

4. An array of useful items—straps (g) or slotted pockets (D)—can be attached to the bands.

5. Also, a slide (C2) can provide solid housing for a ring (k) that is attached to a spring snap-hook (D2).

6. The opposite side of the snap-hook features a rope-clamping jaw (s) that aids in sustaining the suspension of a person at varying altitudes.

In the Inventor's Words

"This invention relates to belts designed to be applied to the person, and provided with hooks and other attachments for securing the person to a fixed object, as also for making an interrupted or uninterrupted descent there from, and for carrying tools and other applications, so as to leave the hands and feet of the user at liberty when working at an elevation or during his ascent and descent."

SMOKE DETECTOR

Patent Name: Smoke Detector
Patent Number: 3,460,124
Patent Date: August 5, 1969
Inventors: Randolph J. Smith, of Anaheim, California,
and Kenneth R. House, of Norwalk, California

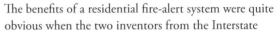

What It Does   The battery-powered smoke detector is an important in-home safety feature, often required by law, which emits a piercing alarm in the presence of smoke, alerting residents of a possible fire.

Background   The benefits of a residential fire-alert system were quite obvious when the two inventors from the Interstate Engineering Corporation received the first battery-powered smoke detector. The prototype used electricity from an outlet and batteries provided a backup source of power.

Today, there are two common types of smoke detector available. One, called an "ion chamber smoke detector," uses a small amount of radioactive material to detect the presence of smoke or heat. This type is more popular as it is less expensive and more sensitive. The other type, described in this patent, uses a photoelectric sensor to detect the changes in light levels caused by smoke particles.

How It Works

Smoke detectors save lives. According to the National Fire Protection Association, homes equipped with smoke alarms have 40-50 percent fewer deaths than homes without them.

1. In figures 1, 2, and 3, a housing (H) includes a main body (6) sitting within a circular wall (14) with apertures (18) in it.
2. A flat circular portion (20) includes an inwards extending socket (22) which supports a lens (24).
3. The lens transmits light from a light source (26) contained within a casing (28).
4. Light travels through the lens, through an inner aperture (32), and into a diffusion and reflection chamber (34).

5. When the chamber becomes filled with smoke, the diffusion of light is disrupted and light is reflected from smoke particles onto a photoelectric cell (50).

6. When this happens, the cell activates an audible alarm device (76) that is connected through a relay.

7. When the alarm is inactive, a ruby glass indicator (64) on the top of the external face of the housing indicates that the light source is working, and the smoke detector is ready and functional.

In the Inventors' Words

"We have provided a smoke detector which is compact in construction and can be mounted conveniently on the wall of a home or other building structure and which is rendered operative merely by placing a battery in the housing and plugging the leads into the transformer (66) in a conventional wall outlet. In its

TIPS FOR EFFECTIVE USE OF A SMOKE DETECTOR

· Install at least one on each level of your home.

· Never remove a good battery or otherwise disable the detectors.

· Check batteries often, and replace when necessary.

· Plan a home escape route in the event of a fire.

· Hold practice drills with your family.

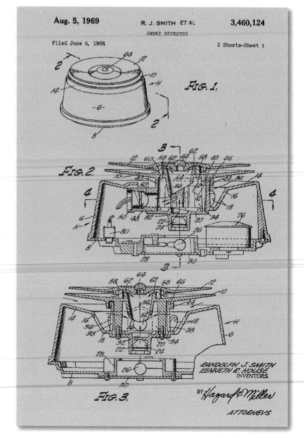

operative condition, light from the lamp (26) is always present to indicate that condition, but the light is not diffused and reflected to the triggering photoelectric cell until there is smoke in the diffusion and reflection chamber (34). The device is extremely reliable. Its operative condition is indicated by illumination of the ruby glass indicator (64) and the alarm being operative either by current from the conventional 115v. service in the home or by the self-contained battery in the alarm itself."

TURTLE EXCLUDER DEVICE

Patent Name: Turtle Excluder Device (TED)
Patent Number: 4,739,574
Patent Date: April 26, 1988
Inventor: Noah J. Saunders, of Biloxi, Mississippi

What It Does   Saves sea turtles caught in nets during the shrimping process.

Background   Little shrimp mean big business—but sometimes shrimping can be bad business for the environment. Shrimp trawling uses a large net, which sweeps a broad swath of ocean space. Often, the net captures more than the intended quarry, including endangered

marine life like the Kemp's Ridley sea turtle. Today, United States environmental policy requires the use of a turtle excluder device (TED) when trawling for shrimp. It is believed that a Georgia fisherman named Sinky Boone invented one of the first TEDs, which was subsequently developed further by the National Marine Fisheries Service (NMFS) between 1978 and 1984. With minimal loss of shrimp, TEDs are effective at excluding up to 97 percent of sea turtles from shrimp nets.

Truly ancient mariners, sea turtles are among the Earth's oldest species, descended from larger ancestors that predate dinosaurs by some 50 million years. Today, the Kemp's Ridley is the most endangered of all sea turtles. But it appears to be in the earliest stages of recovery due to two major factors: protection of nesting females and nests in Mexico, and the requirement of TEDs in shrimp trawls both in the United States and Mexico. In 1996, the U.S. Court of International Trade ruled that all countries that export shrimp to the United States must have an enforcement policy regarding TED usage.

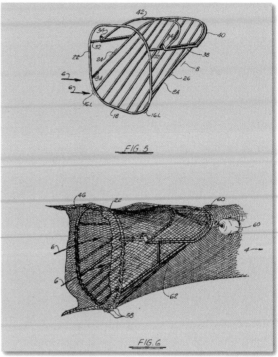

FIG. 5

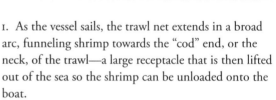

FIG. 6

How It Works

1. As the vessel sails, the trawl net extends in a broad arc, funneling shrimp towards the "cod" end, or the neck, of the trawl—a large receptacle that is then lifted out of the sea so the shrimp can be unloaded onto the boat.

2. Affixed to the trawl at the neck is a grid of parallel bars positioned in a gradual upward angle so that, while shrimp glide right on through, turtles can be maneuvered toward the roof of the net enclosure.

3. Upon contact, the turtle shell activates a release hatch that sends it out of the net through a trapdoor, which closes behind it. The patentee for this invention has introduced an improvement to previous TEDs, intended to reduce shrimp loss when the trap door opens. The trap door is specifically designed to conform to the anatomy of a turtle, allowing no more space than necessary for escape.

Alison Sumner

Kemp's Ridley sea turtle

According to the Shrimp
Council, U.S. consumers
spend more than $10 bil-
lion annually on shrimp.
For developing nations,
trade in seafood prod-
ucts is greater than that
of coffee, tea, rubber, and
banana combined.

"It is the purpose of the current invention to show an improved form for a turtle exclusion device by providing, appended to an angled bar turtle exclusion device, a particular form of trap door escape which is shaped so as to adapt to the shell shape of a turtle, providing for a more ready opening and escape of the sea turtle, and is angled, hinged, and pivoted to the overall excluder so as to provide for a positive closing effect, preventing the loss of shrimp when not actually being activated by an escaping turtle."

VITAMINS

Patent Name: Process for Obtaining Vitamins
Patent Number: 2,049,988
Patent Date: August 4, 1936
Inventor: Robert R. Williams, Roselle and Robert
E. Waterman, of Orange, New Jersey; assignors to
Research Corporation, of New York, New York

What It Does   The process outlined in the patent describes a method of extracting vitamins from their sources to be used as important nutritional supplements.

Background   The word *beriberi* means "I can't, I can't" in Sinhalese. It is also the name of a disease caused by malnutrition and specifically a deficiency of thiamine, once common in Eastern countries, where milled rice was stripped of the thiamin layer on the outside. The effects of the disease can be excruciating, causing severe nerve damage, inflammation, and even death. Robert Williams likely witnessed victims in the throes of this disease as a boy. He was born in India to missionary parents, and he remained there until he was ten.

A wartime nutrition
demonstration for
housewives.

Back in the U.S., Williams attended the University of Chicago, earning a Masters of Science in chemistry. Then he moved to the Philippines to take a job at the Manila Bureau of Science. He spent much of his energies looking for a cure to beriberi. Williams went

A chemist conducts research on the vitamin content of dehydrated vegetables.

In 1936, Merck & Company first introduced thiamin to the commercial market.

back to the States during World War I to work for the Bureau of Science in Washington, DC, and later as the chemical director for the Bell Telephone Company in New York.

Throughout his life, Williams did not forget about the malnutrition suffered by the people from his birth country. Eventually, in 1933, he came up with a way to isolate thiamine in crystalline form. He used what is called fuller's earth—a highly adsorbent claylike substance—to extract the vitamin. Two years later, he could synthesize vitamin B1, the exact vitamin whose deficiency causes beriberi.

His discovery, or rather synthesis, was a medical marvel that helped boost vitamin synthesis and the field of chemically reconstructed vitamins. Williams's accomplishments also contributed to enriching grains in this country to combat the riboflavin deficiency common among impoverished people throughout the world, including in the U.S.

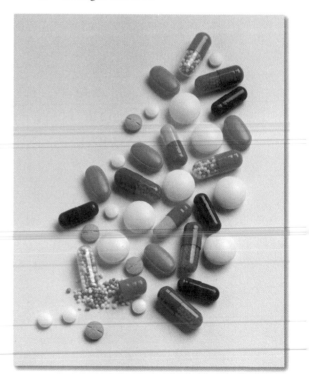

1. Water-soluble vitamins are extracted through an adsorbent filter.
2. The resultant solution is treated and preserved by chemical means.
3. Acyl aromatic chloride, aromatic sulphochloride, or lead acetate can be used to purify the solution.
4. The resulting vitamin solution may be concentrated to a small volume by evaporation in vacuo, or further purified.

In the Inventors' Words

A Federal Enrichment Act passed in the U.S. in 1941 requiring that millers, bakers, and cereal manufacturers must restore the nutritive content to whole wheat products by adding iron, thiamin, riboflavin, and niacin lost in milling or processing.

"According to this invention the vitamin is removed from fuller's earth or similar absorbent by extraction with an excess of aqueous solution of a substance which is strongly absorbed upon the fuller's earth. The solution is preferably used in an acid condition to avoid destruction of the vitamin by heat or alkalinity and is best used at a temperature of 80°C to 100°C to promote complete extraction. The fuller's earth is preferably stirred for several minutes with the hot solution which is then removed while hot by decantation, centrifugation or filtration. The extraction is preferably repeated several times and the extracts combined for further use."

FOUR

The Strange and Sensational Patent

ASTROTURF

Patent Name: Monofilament Ribbon Pile Product
Patent Number: 3,332,828
Patent Date: July 25, 1967
Inventors: James M. Faria, of Decatur, Alabama,
and Robert T. Wright, of Pensacola, Florida

What It Does

Astroturf simulates the look and feel of grass but with added durability, for use both indoors and outdoors for a variety of recreational and sporting activities.

Background

Walt Whitman wrote, "I believe a leaf of grass is no less than the journey-work of the stars." Certainly he could not have anticipated the invention of a factory-made synthetic grasslike turf, which was created after observers in the 1950s noted that city dwellers were less active and less fit than their country cousins. In the early 1960s, Monsanto Industries funded research to test such variables as foot traction, cushioning, and wear resistance in the development of a synthetic fiber that would serve as a viable substitute for grass playing fields. Until then, artificial turf material was developed strictly for decorative purposes. Meanwhile, Judge Roy Hofheinz wanted to build the world's first indoor sports arena. In 1966, the synthetic material, then called Chemgrass, was used to green the playing field of the Astrodome in Houston, Texas. A year later, the invention was patented and the name Astroturf became synonymous with indoor "grass."

Results from a 1995 poll revealed that more than 93 percent of athletes in the National Football League believed that playing on artificial turf increased likelihood of injury.

How It Works

1. Synthetic fibers are woven on a Wilton cut-pile loom to form a woven backing with a cut-pile face extending from one side.
2. A suitable latex formulation on the surface of the backing renders the structure dimensionally stable.
3. A polymeric elastomer is applied to the latex backing to provide stable matting.

In the Inventors' Words

A new generation of Astroturf called AstroPlay® is used for the New York Giants' practice facility.

"The prior art reveals that attempts to make artificial grasses have been made during the past several years. In most instances, the inventive concepts have been concerned primarily with providing a decorative artificial grass and have not attempted to provide an artificial grass which will withstand permanent outdoor installation and the abusive wear caused by spiked or cleated shoes. There are no known simulated turfs which have performance characteristics other than decorative that compare even closely with natural turf."

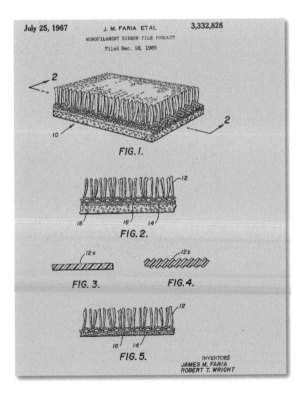

July 25, 1967   J. M. FARIA ET AL   3,332,828

MONOFILAMENT RIBBON PILE PRODUCT

Filed Dec. 28, 1965

FIG. 1.

FIG. 2.

FIG. 3.   FIG. 4.

FIG. 5.

INVENTORS
JAMES M. FARIA
ROBERT T. WRIGHT

BUBBLE WRAP

Patent Name: Method for Making Laminated Cushioning Material
Patent Number: 3,142,599
Patent Date: July 28, 1964
Inventor: Marc A. Chavannes, of Brooklyn, New York, assignor
to Sealed Air Corporation, of Hawthorne, New Jersey.

What It Does

Bubble Wrap is a popular cushioning product used for packing and mailing fragile material throughout the world. (It is also fun to pop.)

Background

An unusual use for Bubble Wrap.

Like so many other invention tales, this one opens with a pair of buddies tinkering around in a garage. In 1957 in Hawthorne, New Jersey, with geek dreams in their hearts and odd materials at their disposal, Marc Chavannes and Al Fielding were trying to make some kind of plastic wallpaper. Instead, they came up with a laminated material that had a lot of bubbles in it. They soon realized its potential as a cushioning product for packaging, and they were off. In 1960, they established Sealed Air Incorporated, which continues to make Bubble Wrap in addition to an impressive line of other products used for packaging. The original material revolutionized the shipping industry—and provided hours of irritating, popping fun to recipients along with their packages.

How It Works

1. Polyethylene resin is extruded through a cylinder with a screw inside that runs its length. As the screw turns, the resin is heated and melted.
2. The resulting liquid is formed into clear plastic film and placed in two stacks.
3. Depicted in patent illustration 9, one layer is wrapped around a drum (10) with circular pockets or other shapes punched in it.
4. Within the drum, vacuum suction draws the film into the holes that form the bubbles.
5. The second layer of film is then laminated, sealing the air inside the holes.

6. Patent illustration 6 depicts the layer (11), with embossed hexagonal pockets (24), which is hermetically sealed to its bottom film layer (16).

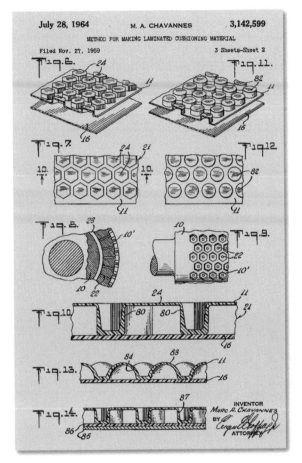

July 28, 1964   M. A. CHAVANNES   3,142,599

METHOD FOR MAKING LAMINATED CUSHIONING MATERIAL

Filed Nov. 27, 1959   3 Sheets-Sheet 2

INVENTOR
MARC A. CHAVANNES
BY
ATTORNEY

Conceived in a garage in New Jersey, the Sealed Air Corporation now includes thirty worldwide labs researching and producing a variety of materials completely unrelated to the original inventors' wallpaper aspirations.

In the Inventors' Words

"Another aspect of the invention resides in an improved method and apparatus for the fabrication of an improved laminated cushioning material wherein at least one lamina or layer is formed to provide a plurality of discrete elements and then a second layer is hermetically sealed to the formed layer to seal the elements and thereby provide sealed pockets in which air or other fluid is entrapped. Since the laminated layers of material have flexibility and elasticity, the resultant laminated material affords a high degree of cushioning as well as shock absorbing action. In the fabrication of such cushioning material, the pneumatic cushioning may be supplemented by an improved mechanical cushioning effected by the design and configuration of the formed elements."

CHIA PET

Patent Name: Animalian figurine
Patent Number: 5,549,500
Patent Date: February 24, 1994
Inventor: Ron Manoah, of Dunwoody, Georgia

What It Does

The Chia Pet is a hollow-bodied decorative ceramic figurine, sculpted with grooves on its external surfaces to accommodate a seed paste. The seed paste sprouts into vegetation to resemble hair on the figurine's surface.

Background

Most Chia Pets get haircuts. These two are longhairs.

While the little Chia Pet may seem nothing more than a last-minute, jokey holiday gift, the practice of growing vegetation as adornment for sculptures is quite ancient. It made its passage from cultural practice to consumer novelty when Joseph Enterprises, Inc. registered the name Chia Pet, and began selling the product in 1982. A registered trademark cannot be used by other companies, but because the basic concept was never patented, others are free to make and sell their own versions.

How It Works

1. An outer covering (11) envelopes an animal figurine (10) made of clay.

2. The covering is scored or textured to retain plant-life nutrient material (13) and a bed of plant seeds (14), while being permeable to liquids and the sprouting of vegetation.

3. Seeds are positioned so that the grass stalks sprouting from them grow through a "scalp" portion to simulate hair.

In the Inventor's Words

"Animal figurines have also been designed which have live herbs that simulate the fur or hair of the particular animal. Exemplary of such is those sold by Joseph

Enterprises, Inc. of San Francisco, Calif. under the trademark Chia Pet. These figurines have hollow, clay bodies in the general form of the animal represented. A large torso portion of the clay body has many small grooves in which moistened chia seeds (Salvia columbariae) are positioned. The moistened chia seeds produce a thick, gel-like paste which binds the seeds to the clay surface. However, the appearance of the seed laden, clay body is unsightly prior to the sprouting of the seeds. Also, because the chia sprouts cannot draw nutrients from the hardened clay body they quickly die and become withered and unsightly. Furthermore, the paste-like substance produced by the seeds is susceptible to causing stains upon contact."

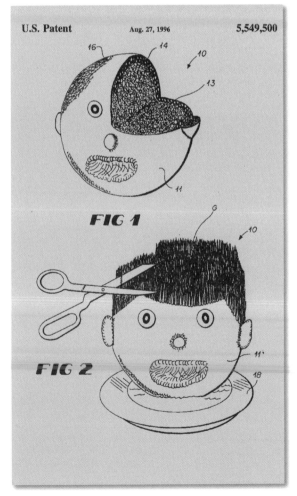

U.S. Patent   Aug. 27, 1996   5,549,500

FIG 1

FIG 2

CHICKEN GOGGLES

Patent Name: Eye-Protector for Chickens
Patent Number: 730,918
Patent Date: June 16, 1903
Inventor: Andrew Jackson, Jr., of Munich, Tennessee

What It Does

When chickens are cooped, they tend to attack one another. Eye protection helps to prevent them from being blinded by the pecking. (The goggles are not intended as a vision aid for fowl.)

Background

This seemingly outlandish invention actually had a very practical purpose: to protect chickens from one another. When chickens are cooped together and one of them sustains some sort of blood-shedding injury, its fellow chickens tend to go into a frenzy. If one chicken has blood on its plumage, its neighbor will attack, spattering more blood and igniting a chain reaction of chicken terror. The birds use their pointed beaks as weapons, oftentimes blinding, maiming, and killing one another. Many small chicken farm operations at the turn of the twentieth century had their own methods of safeguarding their fowl, and eye protection was not uncommon.

In the 1930s, a company began marketing chicken glasses with rose-colored lenses. The reddish tint camouflaged the sight of the blood that tended to drive the chickens nuts. Today, most large chicken and poultry operations keep their birds separated. At free-range chicken farms, the pointed beaks of chicks are snipped at an early age to prevent carnage and mutilation of the poultry. But you can't help loving this chicken eyewear, which lent a certain professorial aura to the bird, belying its innate bloodlust and brutality.

How It Works

Resembling a miniature pair of Lennon specs, this invention consists of two circular frames connected by a loop that fits over the chicken's beak. A U-shaped strap attached to the loop where it meets the circular frames fastens around the chicken's head.

"This invention relates to eye-protectors, and more particularly to eye-protectors designed for fowls, so that they may be protected from other fowls that might attempt to peck them, a further object of the invention being to provide a construction which may be easily and quickly applied and which will not interfere with the sight of the fowl."

ESCAPABLE COFFIN

Patent Name: Improved Burial Case
Patent Number: 81,437
Patent Date: August 25, 1868
Inventor: Franz Vester, of Newark, New Jersey

"The boundaries which divide Life from Death are at best shadowy and vague. Who shall say where the one ends, and where the other begins? We know that there are diseases in which occur total cessations of all the apparent functions of vitality, and yet in which these cessations are merely suspensions, properly so called."
—Edgar Allan Poe, "The Premature Burial"

What It Does

Provides a means of escape for anyone finding himself in the unfortunate circumstance of having been buried alive.

Background

Being buried alive certainly sounds like the macabre stuff of Poe, but such misfortunes were probably not completely uncommon prior to the twentieth century. Because medical means for determining death were less sophisticated than they are today, coma victims, for example, ran the risk of being sent to their graves prematurely. Franz Vester's invention was not the first in a series of "safety coffins" with escape hatches or signaling means, nor was it the last. But, certainly it was among the most elaborate.

How It Works

1. The square tube (C) extends from above the surface of the grave to a base (D) on the lid (B) of the coffin; and includes air-inlet openings (F) that connect into the coffin.

"What would Mozart be doing if he were alive today? Scratching at the top of his coffin!" Not so, if he had been buried in Franz Vester's "improved burial case," patented in 1868.

2. At the base on the lid is a sliding glass door (L), held in place by a spring bar (E), which allows investigators, or the morbidly curious, a peak at the face inside the casket from above ground.

3. Inside the tube is a ladder (H), and on top of the tube is a bell (I).

4. Attached to the bell is a cord (K) which extends the length of the tube and into the coffin, where its end is laid in the buried person's hand.

5. Inside the coffin lid, the sliding glass door is ready to be closed by a spring (M) activated once the tube is withdrawn—when the deceased is determined to be, with unwavering certainty, deceased and ready, at last, to rest in peace. 9.

6. When the glass door shuts, air is closed off from the coffin; and the tube can be reused.

7. Ashes to ashes, dust to dust; if death be uncertain, this coffin is a must.

In the Inventor's Words

Death be not proud. Or cheap. The average casket sold in the United States now costs slightly more than $2,000; some sell for as much as $10,000, according to the Federal Trade Commission.

"The nature of this invention consists in placing on the lid of the coffin, and directly over the face of the body laid therein, a square tube which extends from the coffin up through and over the surface of the grave, said tube containing a ladder and a cord, said cord being placed in the hand of the person laid in the coffin, and the other end of said cord being attached to a bell on the top of the square tube, so that should a person be interred ere life is extinct, he can on recovery to consciousness, ascend from the grave and the coffin by the ladder; or, if not able to ascend by said ladder, ring the bell, thereby giving an alarm, and thus save himself from premature burial and death; and if, on inspection, life is extinct, the tube is withdrawn, the sliding door closed, and the tube used for a similar purpose…"

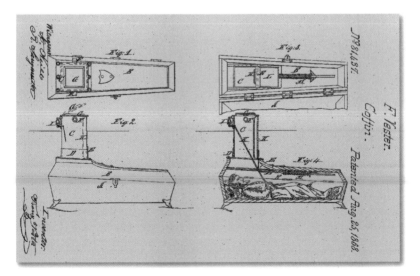

GENETIC ENGINEERING

Patent Name: Process for producing biologically
functional molecular chimeras
Patent Number: 4,237,224
Patent Date: December 2, 1980
Inventors: Stanley Cohen, of Portola Valley, California,
and Herbert Boyer, of Mill Valley, California

What It Does   Manipulates DNA structure by incorporating replicable genes into cells.

Background   The science of genetics is the science of heredity and
of the mechanics by which characteristics are passed
from one generation to the next, whether from virus to

A British model-builder's
concept of genetically
engineered humans.

virus, fruit fly to fruit fly, or obnoxious parent
to obnoxious child. The seed for the concepts
of genetics was planted by the Augustinian
monk Gregor Mendel in 1861, when he showed
that hereditary traits in pea plants were transmitted from one generation to the next in pre-
dictable patterns.

Much later it was shown that Mendel's laws
applied to all species, including humans. However, in many cases the inheritance of genes turned out
to be much more complicated than Mendel had theo-
rized. The inheritance of complex traits was shown to
be due to the interaction of several Mendelian factors
or genes.

A Finn Dorset sheep was
born in July 1996. She
was named Dolly after
country singer Dolly Par-
ton and she became just
as famous as her name-
sake. She was cloned
from the breast cell of an
adult ewe. She matured
and mated naturally with
a mountain ram. Dolly
gave birth to a healthy
lamb and to the ongoing
debate on the ethics of
human cloning.

In the 1940s experiments with bacteria established
that deoxyribonucleic acid, or DNA, is the chemical
basis of the gene. Enter James Watson and Francis
Crick, who are credited with discovering the structural
basis of genetic information in DNA and opening
the floodgates in the field of genetics and the many
branches of the science that have flourished from it.

In 1953, James Watson and Francis Crick described
the structural basis of genetic information in DNA as a
double helix. This structure resembles a ladder twisted
in a clockwise spiral. The order of nucleotides (the
crossbars of the double helix) along the DNA molecule
was found to specify the linear arrangement of amino

acids in proteins. Thus, the genetic code governs the translation from nucleotide order to amino acids to proteins to the very structure of living organisms as far apart as viruses and humans.

The genetic code functions by the rule that triplets of DNA nucleotides each specify a single amino acid. Restriction enzymes also each specify a single amino acid. Restriction enzymes cut DNA at specific sites and allow the in vitro recombination of DNA within and between organisms and between species. These and other discoveries led to the development of molecular biology, and eventually to genetic engineering.

By early 1973, Herbert Boyer and Stanley Cohen added replicable genes to a simple cell, giving birth to recombinant DNA. Their findings have not only been of great fundamental significance but are increasingly applied in many branches of genetic biology including agriculture, animal breeding (hello, Dolly!), and medicine. Genetic engineering, or gene manipulation, is now a widely debated experimental technology developed to alter the genome of a living cell for medical or industrial use. Indeed, the method described in this patent has opened the doors to a strange new world.

How It Works

1. Plasmids or virus DNA are cleaved to provide linear DNA having ligatable termini.
2. A gene possessing complementary termini is inserted, providing a biologically functional replicon with a desired phenotypical property.
3. The replicon is then inserted into a microorganism cell by transformation.
4. Isolation of the transformants provides cells for replication and expression of the DNA molecules present in the modified plasmid.

In the Inventors' Words

"The ability of genes derived from totally different biological classes to replicate and be expressed in a particular microorganism permits the attainment of interspecies genetic recombination. Thus, it becomes practical to introduce into a particular microorganism, genes specifying such metabolic or synthetic functions as nitrogen fixation, photosynthesis, antibiotic production, hormone synthesis, protein synthesis, e.g. enzymes or antibodies, or the like—functions which are indigenous to other classes of organisms—by linking the foreign genes to a particular plasmid or viral replicon."

GOAT TRACTION FOOTWEAR

Patent Name: Footwear with mountain goat traction elements
Patent Number: 6,226,896
Patent Date: May 8, 2001
Inventor: Michael Ray Friton, of Portland, Oregon; assigned by Nike, Inc.

What It Does
The shoe invention described in this patent takes as its model the impressively sure-footed mountain goat..

Background

One of the most adept mountain climbers in North America, the desert bighorn sheep is now trying to climb its way off of the Endangered Species list.

A former University of Oregon college runner named Phil Knight went into the sneaker business in 1962. He adapted a waffle-shaped sole, a cushioned midsole, a wedged heel, and nylon tops, outstripping his predecessors in the sport shoe business with the lightness and flexibility of his product. A decade later, he was marketing shoes named after the Greek goddess of victory, Nike.

Because of the growing interest in trekking the outdoors, an array of hiking shoes and bulky boots began lining the shelves of shoe stores worldwide in the eighties. Lightweight trail shoes and durable sandals soon followed, including the Nike Mada and Terra trail shoes and the Nike Terra and Deschutz.

How It Works
1. The sole has a ground-engaging surface including soft, compliant traction elements and one or more relatively hard lugs.
2. The lugs are stiffer in compression than the traction elements; and run adjacent to them.
3. The traction elements extend downwardly below the lugs such that, in use, a bottom surface of the traction elements will make initial ground contact and partially compress.

4. The compression cushions the impact of the ground engagement and increases the ground contact. The bottom surface of the lugs touches the ground.

5. The lugs limit compression of the traction elements and serve as a relatively rigid catch for irregular and soft ground surfaces.

6. Cushioning is attached to the upper.

7. The sole has a ground engaging surface including an outer perimeter border and an interior.

8. The interior region comprises a group of relatively soft traction elements.

9. The border region comprises a pair of relatively hard lugs, stiffer and adjacent to the group of traction elements.

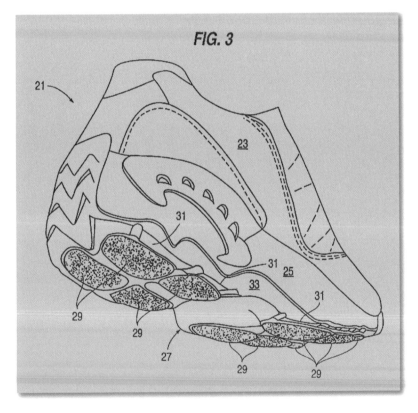

FIG. 3

In the Inventor's Words

"The effectiveness of a mountain goat's hoof in providing that animal with sure footing on diverse and extreme ground conditions has been recognized. As described in the book entitled *Beast the Color of Winter, the Mountain Goat Observed*, by Douglas H.

Chadwick, Sierra Club pub. (1983), '[t]he sides of a mountain goat's toes consist of the same hard keratin found on the hoof of a horse or deer. Each of the two wrap-around toenails can be used to catch and hold to a crack or tiny knob of rock ... The mountain goat is shod with a special traction pad which protrudes slightly past the nail. This pad has a rough textured surface that provides a considerable amount of extra friction on smooth rock and ice. Yet it is pliant enough for any irregularities in a stone substrate to become impressed in it and thereby add to the skid proofing effect.'"

KITTY LITTER

Patent Name: Cat box filler with incorporated pesticide
Patent Number: 4,664,064
Patent Date: May 12, 1987
Inventor: Edward H. Lowe, of Cassopolis, Michigan

"No matter how much cats fight, there always seems to be plenty of kittens."
—Abraham Lincoln

What It Does
Kitty (or cat) litter is an absorbent, easy to replace substance resembling sand, used in a box or bin as a place for housecats to do their business. The litter keeps odors and messes to a minimum, and cats seem to appreciate having their own neat and tidy toilet facility.

Background
This patent is but one of dozens awarded to Ed Lowe who has earned the odd designation as "the father of kitty litter." Lowe's own father Henry was a small

business owner who sold industrial absorbents in 1948. Fresh out of the Navy, Ed joined Dad in the biz and tried in vain to sell farmers a new "Fuller's Earth" product as a nesting material made from partially oven-fired balls of dried clay. While local farmers did not exhibit much interest, Ed's next-door neighbor did. In response to her cat-related woes, Ed suggested

With the introduction
of commercial cat litter
products, the cat became
more of an indoor pet.
Keeping your cat indoors
can:

· Reduce the spread of
disease, such as rabies.

· Save birds, mice, and
lizards, which all natu-
rally control insect pop-
ulations.

· Reduce its chance of
getting mauled by the
neighbor's German
Shepherd or tortured by
the local delinquent.

using the absorbent material. She was very impressed
by the results and kitty litter was born.

Until kitty litter marked its territory in the commercial market and established the beginnings of a
new industry, cat owners commonly used shredded
newspaper, dirt, or sand for in-home litter. In the
meantime, Lowe became an ambitious traveling salesman and hit the road aggressively hawking his kitty
litter. As his product began to absorb public attention
and sales, Lowe established Edward Lowe Industries,
Inc. and concentrated his energies on making improvements, such as added ingredients to kill odor-
causing bacteria or to create moisture-activated clump-
ing action for easier cleaning. This patent represents a
litter that contains pest-controlling chemicals, helping
to keep your cat flea-free while it makes pee. The kitty
litter king died in 1995.

How It Works

Cats are very meticu-
lous, constantly cleaning
themselves. Their rough-
surfaced tongues are
designed to lick off any
scents that may attract
insects or predators.

1. The pesticide may be either vapor producing or
contact-actuated.
2. A contact-actuated pesticide is impregnated into
clay, or any solid, non-toxic organic polymer.
3. The product should be packaged either directly
with the cat box filler or in a separate pouch to ensure
containment.
4. When the cat scratches in its litter box, it releases
the pest-control chemicals.
5. Examples of contact-actuated pesticides are hep-
tenophos, carbaryl, propoxur, resmethrin, naled,
trichlorfon, tetrachlorvinphos, diazinon, and fenchlor-
phos.

In the Inventor's Words

"Pest control in domestic animals has long been a
major concern of pet owners. Previously, adequate
pest control has been achieved by the use of pesticide-
impregnated collar worn by the animal. The major
drawback of these collars is that they must be worn
constantly by the animal to be effective. This constant
wear increases the risk of pet poisoning and presents a
danger of accidental strangulation in cats.

"The pest control of this invention involves the in-
corporation of a pesticide into granules which are dispersed in the cat litter box, when the cat uses the litter
box, the pesticides granules bathe the cat in vapor and

prevent the outbreak of fleas and other pests. Because the granules are dispersed within the cat box filler, pest control is achieved in the litter box as well as on the cat itself."

LAVA LAMP

Patent Name: Display Device
Patent Number: 3,570,156
Patent Date: March 16, 1971
Inventor: Edward C. Walker of Hampshire, UK

What It Does
The lava lamp is an ornamental lamp consisting of a transparent glass vessel and a visually pleasing lava-like substance that continually bubbles to the surface and then flows back down when the lamp is on.

Background
Edward Craven Walker once described the spectacle of his invention as a sexy thing that, "starts from nothing, grows possibly a little bit feminine, then a little bit masculine, then breaks up and has children." A World War II veteran, Walker adopted the lingo and the lifestyle of the flower children. Part Thomas Edison, part Austin Powers, he was a nudist in the psychedelic days of the UK—and possessed some pretty savvy marketing skills to boot. "If you buy my lamp, you won't need to buy drugs," he was known to say.

The world's largest lava lamp

In the mid-sixties Walker hit on the idea of using a lamp concealed within a base to display a groovy glass vessel that contained a lavalike oil and wax substance swimming in a colored liquid. As the lamp heated the substance in the globe, the lavalike substance would change form and shape. At first, shopkeepers resisted the strange lamp. But the lava lamp was soon illuminating nightclubs and flats across London.

Eventually, a pair of enterprising Americans bought rights to manufacture the lava lamp in the U.S. Their company slapped on the Lava Lite® trademark, and the lamps have been selling continuously ever since

How It Works

1. Dyed water and another liquid of mineral oil, paraffin wax, paraffin, and carbon tetrachloride is sealed into a glass container.

2. The container rests in a hollow base which contains an electric light bulb.

3. When the lamp is off and cold, the second liquid congeals in a hard lump at the bottom of the glass.

4. When the lamp is on, the substance expands, loses density, and rises.

5. As it rises, it moves further from the heat source and begins to resolidify and sink.

6. This heating and cooling process is constantly reshaping the substance swimming in the dyed liquid as it is illuminated by the lamp.

In the Inventor's Words

"I think it will always be popular," Walker told the Associated Press before his death in 2000. "It's like the cycle of life. It grows, breaks up, falls down, and then starts all over again."

Apparently, random numbers aren't as easy to come by as you and I would think. Lavarand, a project underway at Silicon Graphics, is using the unpredictable motion of lava lamps to help generate data like this:

Ü·}7ÀÈ»o @¹3?s+ïŸ¨v»
HµÎ́œ³k tü; Ãµî¨ wå
Z ÕïJ²0@÷q'¦Å"þ ü,
d£?C@| &§zÑxØÂ½Jß-
y#poz?‰oÙ ⊠-Ð
â¨^S£GMþ‡ w¸¢3 Ý
}ÞE|?ÅÃxCù ¥ xDÉw2
=„"Û^ün^îèå,ÄO> ˆ„q8É«
?1jó¿Èÿ£Ñæs§j©Œ*#Ó
w_?6

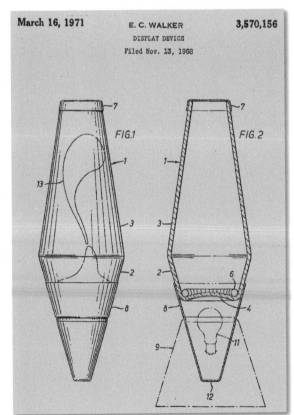

March 16, 1971   E. C. WALKER   3,570,156
DISPLAY DEVICE
Filed Nov. 13, 1968

FIG.1   FIG.2

NEON

Patent Name: System of Illuminating by Luminescent Tubes
Patent Number: 1,125,476
Patent Date: January 19, 1915
Inventor: Georges Claude, of Paris, France

What It Does   Neon gas fills tubes to provide a cool, long-lasting, oddly brilliant, nonincandescent lighting tailor-made for signs that must be seen from a long distance.

Background   Neon is a rare, inert gas found in very small quantities in the air. In the late eighteenth century, scientists used glass tubes to observe the results of experiments that involved high-voltage charges. After the five noble gases—helium, neon, krypton, argon, and xenon—were discovered in the 1890s, Georges Claude observed that electrical discharges passing through some of these gases could produce light without incandescence. Neon was the gas that produced the brightest light.

In 1910, Claude introduced the first neon light display in Paris. In 1923, the French scientist-turned-entrepreneur sold two neon signs to a Los Angeles car dealership, and within a year, he owned a chain of "Claude Neon" stores throughout the United States. As the neon lighting industry grew more sophisticated, neon signs became a regular feature of the urban landscape, adorning movie theaters, car dealerships, bowling alleys, restaurants, and bars.

Viva Las Vegas. Las Vegas is the ultimate neon city, luring tourists and their dollars from all over the world to see what has been called the first truly postmodern city in the world.

This patent was awarded to Claude five years after his first French patent for neon lighting and represents a refinement in his process. He included methods in this patent to obtain a more brilliant neon light, and a way to keep gaseous deposits at the electrodes from altering the color of the tube.

How It Works   1. A vacuum pump (c) removes air from a glass tube bent into the desired shape.
2. Another tube (d) inserts the neon gas.

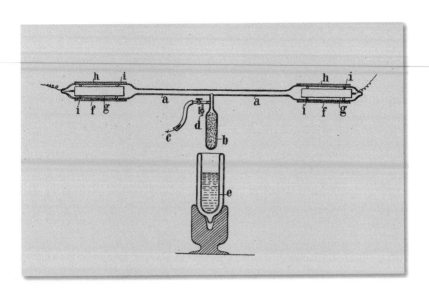

3. A carbon receptacle (b) is immersed in liquid air (e) to facilitate removing excess foreign gases and give the tube its fullest possible brilliancy.

4. The lamp is then separated from the carbon receptacle.

5. The tube (a) contains electrodes (g) at either side to receive electrical current and illuminate the gas within the tube.

6. When a neon lamp is plugged into a source of electricity and turned on, the gas becomes illuminated.

In 1912, the world's first neon sign illuminated the words "Le Palace Coiffeur" above a Paris barbershop. In that same year, Cinzano Italian vermouth became the first product advertised by neon lighting.

In the Inventor's Words

"What I claim and desire to secure by Letters Patent of the United States is … (a) luminescent tube containing previously purified neon and provided with internal electrodes for illuminating said gas, said electrodes being deprived of occluded gases and having an area exceeding 1.5 square decimeters per ampere, to decrease the vaporization of the electrodes and prevent the consequent formation upon the walls of the tube, in proximity to said electrodes, of deposits containing said gas, whereby the luminosity of the tube is maintained constant for a very considerable period of time without a fresh introduction of gas."

OUIJA BOARD

Patent Name: Toy or Game
Patent Number: 446,054
Patent Date: November 10, 1891
Inventor: Elijah J. Bond, of Baltimore, Maryland;
assignor to Charles W. Kennard and William H. A.
Maupin, both also of Baltimore, Maryland

What It Does The Ouija board is a novelty used for conducting
séances and communicating with the dead.

Background According to the Museum of Talking Boards, the year
1848—the year before Edgar Allan Poe died—was
pivotal in the rise of modern Spiritualism. That year,
in a New York cabin, two sisters contacted
the spirit of a dead person and started a
fervor for the occult in the United States
and Europe. Spiritual mediums, crystal-ball
gazers, palm readers, and a motley crew of
others rose like zombies from the grave.
One popular method for conversing with
"the other side" was through table-turn-

Before there were
television psychics, there
was the Ouija.

ing. A specialist would conduct a séance in
which attendees placed their fingers on a table edge;
upon contact, the table might tip in one direction or
the other as a mounted dial-like mechanism pointed to
letters and spelled out messages. The stage was set for
the introduction of the Ouija board.

Elijah Bond was acquainted with a shrewd group
of men who were no doubt aware of the public interest in the occult; they lived in the very city where Poe
was buried. Certainly, they concluded, there was more
of a market for the supernatural than just books and
carnival hoodoo gimmickry. The Kennard Novelty
Company was established a year before this patent was
issued and became the first company to commercially
manufacture the Ouija Board. Its precise cultural origin remains a subject of debate, but the board's promise of a means to communicate with the underworld
was very attractive to many, in spite of the fact that the

patent itself clearly characterizes the item as a "Toy or Game."

After company power struggles, ownership of the Ouija board product changed hands several times and legal battles ensued. The owners apparently lost a battle with the Internal Revenue Service when they tried to claim that their product was a spiritual tool and, therefore, its sales should not be taxed. In the late 1960s, a new age of spiritualism sparked new interest in, and even fear of, the talking board. It was then, too, that ownership of the Ouija board went to Parker Brothers (now a subsidiary of Hasbro), based in Salem, Massachusetts, where a number of burned witches probably still have a lot to get off of their charred chests.

Is there really life after death? Yes, at least for the Ouija. Today, the Parker Brothers Company sells Ouija boards that are reproductions of the spirit-summoning board sold more than one hundred years ago. They hold all of the patents and trademarks to the board and they sell lots of them. How many? You'll have to ask the Ouija.

To Infinity and beyond. You can talk to spirits in cyberspace with an online Ouija board at many Web sites.

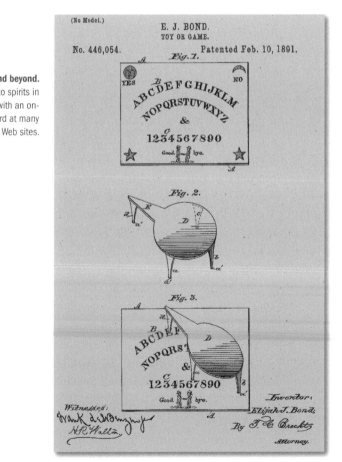

How It Works   A flat-surfaced board approximately 15 by 22 inches in size features the letters of the alphabet in two arching rows, the symbol "&" below the letters, and below this, the digits 1, 2, 3, 4, 5, 6, 7, 8, 9, and 0 straight across. Also printed on the board are the words "YES," "NO," and "GOODBYE" with accompanying illustrations. A stout table-like structure (D) features an elongated section (E). The little table is placed on the board. The hand of the player is placed lightly on the table. Questions are posed. The leg beneath the elongated portion of the table will point to letters that may be interpreted as answers.

In the Inventor's Words   "My invention relates to improvements in toys or games, which I designate as an "Ouija or Egyptian luck-board;" and the objects of the invention are to produce a toy or game by which two or more persons can amuse themselves by asking questions of any kind and having them answered by the touch of the hand, so that the answers are designated as letters on a board."

PIGEON STARTER

Patent Name: Improvement in Pigeon Starters
Patent Number: 159,846
Patent Date: February 16, 1875
Inventor: Henry A. Rosenthal, of Brooklyn, New York

What It Does   Startles pigeons from traps for an improved recreational shooting experience.

Rothschild Petersen
Patent Model Museum

A wooden predator realistic enough to fool some pigeons all of time.

Background   In the nineteenth century, sportsmen enjoyed target-shooting with live pigeons, which were often kept in traps. When their owners wished to take target practice, they would bring the cage to an open field and unlatch the door. The problem was that pigeons were often reluctant to fly out of the trap; either, the birds were not too bright, or on the contrary, they possessed an intuition as to what this new freedom really might mean. Enter the pigeon starter—a likeness of a natural predator meant to startle the pigeon out of the trap and into flight.

The pigeon starter is today no more than an historic curiosity. With automated clay trap dispensers, trap shooting remains a popular sport without contributing to any loss of avian life.

How It Works

There's a story of an English sportsman who, during an especially frustrating day afield, decided to tie one of his pigeons to a stool, thereby making it a live decoy to attract other pigeons. Hence, the term "stool pigeon" refers to a police informer, a person who can bring trouble near.

1. A figure of a cat or other animal (A) includes four rigid legs that are pivoted at their upper ends.
2. The feet of the animal are attached securely to a baseboard (B) designed to be staked to the ground.
3. Between the fore and rear legs, a coiled spring (C) is secured to the baseboard and is attached to the lower end of a lever (D).
4. The upper end of the lever is secured to the underside of the animal's body.
5. A catch (E) is attached to the base and held forward by a spring (F) also attached to the baseboard in order to hold the lever when the animal is pressed down.
6. A cord (G) is attached to the end of the spring catch (E, F) which, when pulled, activates the mecha-

According to the Smithsonian Institution, the passenger pigeon once constituted 25 to 40 percent of the total bird population of the United States. At the time Europeans first came to America, there were an estimated 3 to 5 billion passenger pigeons. Overhunting led to the decline of the species, and at the time of this patent, it was in serious peril. On September 1, 1914, the species became officially extinct when the last known passenger pigeon died at the Cincinnati Zoo.

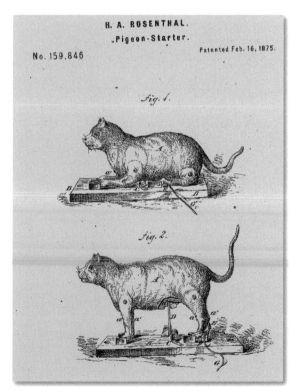

H. A. ROSENTHAL.
Pigeon-Starter.

No. 159,846

Patented Feb. 16, 1875.

Fig. 1.

Fig. 2.

nism and abruptly launches the animal likeness into an erect position.

In the Inventor's Words "As traps have heretofore been used, it frequently happens that the pigeons will not leave the trap when it is sprung, and have to be frightened out by shouting or throwing stones, &c., which tends to make the sportsman nervous, and frequently causes him to lose his shot."

This pigeon starter is formed by "the combination of the body, the rigid pivoted legs, the spring, the lever, the spring-catch, and the trip-cord with each other and a base-board; and in the employment, as a pigeon-starter, of a figure made to represent the body of an animal, and provided with suitable mechanism to enable it to be set in a crouching position, and to be sprung into an erect position …"

THEREMIN

Patent Name: Method of and Apparatus for the Generation of Sound
Patent Number: 1,661,058
Patent Date: February 28, 1929
Inventor: Leon Sergejewitsch Theremin, of Leningrad, Russia

What It Does This electronic music-making device emits haunting tones when "played" by moving one's hands in the vicinity of its active parts, two radio frequency oscillators.

Leon Theremin with his invention, 1928.

Background Leon Theremin was born Lev Terman in Russia in 1896. One day, he was putting together parts for a radio when he began hearing noises he knew were not part of any radio broadcast. He soon realized he could manipulate and control the frequencies of the sounds by waving his hands near the device. From these initial observations of the interruption of sound waves, Terman invented the world's first electronic musical instrument, aptly named the "theremin," in

1919. The invention was first publicly demonstrated at the Moscow Industrial Fair in 1920. Lenin himself commissioned six hundred of the instruments to be built and toured around the Soviet Union.

Theremin (as he renamed himself) emigrated to the United States and soon succeeded in showcasing his musical ingenuity. Audiences everywhere were amazed as he played the theremin. Here was a musical instrument that was played without physical contact, by manipulating the energy currents in the air around it with the movement of the hands: one hand controlled volume; one controlled pitch. The result was an eerie, violin-like sound with a haunting beauty and magical quality. The invention was a marvel of both science and art, and the earliest example of electronic music.

In addition to the Beach Boys' use of the theremin in "Good Vibrations," Led Zeppelin, a hugely popular band in the '70s, was known to use a Theremin from time to time. More recently, the Jon Spencer Blues Explosion featured the instrument in the 1994 album *Orange*, and even featured a rendering of the instrument on its album cover.

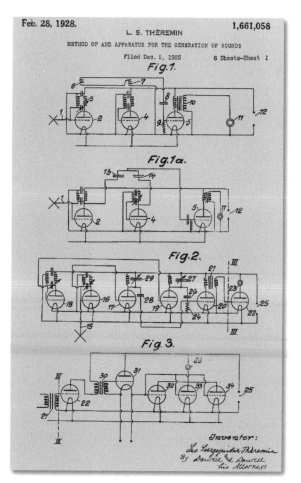

Today the sounds of the theremin are recognizable to many as the high-pitched quavering notes used to score sci-fi flicks of the fifties, and used by the rock group the Beach Boys, most notably in their song, "Good Vibrations." But the sound of the theremin is no more bizarre than the history of the man who created it.

The inventor settled in New York City, living large with lots of friends and admirers. He went on to develop a variation of the instrument that could be played by the movement of the human body, so that dance motions could be translated into music. There was even a troop of "theremin dancers," as they were called. In 1938, he suddenly and inexplicably disappeared, only to reappear a half century later.

Theremin had been kidnapped from his New York apartment and taken to the Soviet Union, where he was made to work on a variety of secret government projects, including the development of the "bug" for the KGB. Eventually, in his old age, he reunited with some of his friends from the United States and enjoyed the recognition of his contributions to electronic music. He died in Moscow in 1993, at the age of 97.

A 1993 documentary film, *Theremin: an Electronic Odyssey*, offers great insight into the inventor's strange and fascinating life. In it, Robert Moog, who went on to create the Moog Synthesizer and other innovations in electronic sound, describes his own early interest in the theremin, dozens of which he made as a kid.

How It Works

The theremin is by no means an easy instrument to play; there are no specific reference points for the musician to read at all. Sound is produced via radio-frequency oscillators. One oscillator is set at a fixed frequency; another at a variable frequency. One hand moves around the metal loop controlling volume. The other hand controls pitch by adjusting its proximity to the antennae.

In the Inventor's Words

"An instrument embodying the invention comprises a sound reproducer, such as a telephone receiver or loud speaker, connected to an oscillating system adapted to be controlled or affected by an object or objects, such as the hands or fingers of an operator held in relative position in proximity to an element of the system."

"TOWERING INFERNO" FIRE BLANKET

Patent Name: Apparatus for Extinguishing Fires in High Rise
Block Buildings or Uniform Transverse Cross-Section Plan
Patent Number: Great Britain 1,453,920
Patent Date: October 27, 1976
Inventor: Arthur Paul Pedrick, of Selsey, Sussex, United Kingdom

What It Does
This innovation in fire control was intended to provide an alternative method to extinguish fires in tall buildings, thus minimizing the casualties and injuries in a tragic "towering inferno" situation.

Background
Disaster movies were big in the early 1970s, and the 1974 release of *The Towering Inferno*, starring Steve McQueen and Shelley Winters, among other luminaries, was the king of the genre. The trend may have triggered the imaginative spark of inventor Arthur Paul Pedrick, who set about devising a disaster-aversion scheme. Listing himself as the subject of a "One-Man Think-Tank" in this British patent application, Arthur Pedrick describes one of his many intriguing inventions: a giant fire-resistant blanket that rolls down from the top of a tall building to smother a fiery conflagration.

How It Works
1. Spindles are secured to the rooftop of a building.
2. Around them are spun rolls of fire-resistant material with weights at their ends.
3. In the case of a fire, the curtains may be released manually or by a thermo-responsive switch.
4. Guide ducts secured along the building may be used to help align the weights and draw the curtain down steadily.
5. To avoid the potential of suffocating people inside the building, the blanket includes strategically placed holes that align with windows in designated "fire emergency" rooms.

In the Inventor's Words
"It is generally impracticable to construct fire escape ladders of over 200 ft. in height... (W)hen fires have occurred in high rise buildings, people cut off from the

lifts or fire escape ladders by the fires starting below them have been forced to jump for their lives, if they are not rescued in time by a helicopter landing on the roof."

Some of Pedrick's many patent applications:

· Earth Orbital Bombs as Nuclear Deterrents

· Reducing the Tendency of a Golf Ball to Slice or Hook by Electrostatic Forces

· Automobiles Driven from the Back Seat

· Aerial Ships Supported by Vacuum Balls or Other Forms of Evacuated Vessels

· Speed of Light Regulated Clock

· Photon Push-Pull Radiation Detector for Use in Chromatically Selective Cat Flap Control

· 1000 Megaton Earth-Orbital Peace-Keeping Bomb

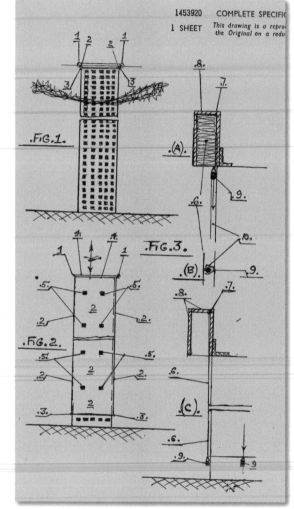

ZEPPELIN

Patent Name: Navigable Balloon
Patent Number: 621,195
Patent Date: March 14, 1899
Inventor: Ferdinand Graf Zeppelin, of Stuttgart, Germany

What It Does

The zeppelin dirigible is a lighter-than-air craft with a rigid frame designed as a vehicle for passengers or freight.

Background

Long before the Wright Brothers took flight in 1903, another pair of brothers achieved a feat of similar magnitude. The French Montgolfier brothers demonstrated the first flight of a model hot-air balloon in 1783. Others in France that year were quick to follow, and human flight was, in effect, achieved. Over time, different styles and models of balloonlike aircraft were conceived, including dirigibles, which were self-propelled lighter-than-air craft and featured directional control surfaces. One inventor of such airships is now permanently linked by name to his invention: Ferdinand Graf Zeppelin.

Zeppelin dirigibles.

In 1997 the Zeppelin Company built a new airship that is certified for passenger flight.

A former Prussian military officer, Zeppelin built his first rigid dirigible—composed of a sturdy lightweight framework—in 1900. Filled with hydrogen gas, the ship was approximately 420 feet long and 40 feet in diameter. A pair of gondolas carried the passengers and crew. Rudders at the front and back provided steering, and an internal combustion engine was used to propel it.

But Zeppelin had bigger ideas: fleets of the large airships moving great quantities of freight. In 1908, he created the Zeppelin Foundation to further his ideas in aerial navigation. Two years later, one of his dirigibles was licensed to provide the first commercial passenger air service of its kind.

The largest airship made by Zeppelin's company was named the Hindenburg. It is said to have been designed to reflect the greatness of Nazi Germany.

Stretching more than 800 feet—the length of several city blocks—the luxury passenger craft offered two-day trips from Europe to North America. The Hindenburg would later be compared to the *Titanic*, not only for its size but for its tragic fate. On May 6, 1937, in Lakehurst, New York, the Hindenburg, carrying passengers from Europe, burst into flames just as it was docking, and thirty-five people died in the tragedy. The event marked the abrupt end of the golden era of the passenger airship, and has lingered in memory as it was captured dramatically on film.

How It Works

A framework is divided into separate compartments. A main gasbag and auxiliary gas bags are contained within each compartment. A moveable carriage supports the weight with rotary drums and also attaches to the rope and the gondolas. A series of balloons coupled together are provided with rigid casings, the front casing being provided with a driving mechanism. With coverings securing the casings and the spaces between the paired balloons, the remainder of the aircraft is adapted to carry the load or freight.

A 1969 ground-breaking rock-and-roll debut album featured on its cover the ominous and shadowy underside of the Hindenburg as its back end exploded in flame. The album, like the band, was called Led Zeppelin.

1. A running weight (b) is suspended by pulley ropes and tackle (b^1, b^2) beneath the ship.
2. Two gondolas (g) are suspended beneath the ship as well.
3. The pulley ropes are accessible from the gondola to manipulate the weight to maintain the ship's balance as it rises or moves forward.
4. Also suspended beneath the ship, a gangway (l) extends almost the length of the entire ship.
5. Operators have full access to the ship by rope ladders (f) to make adjustments to air.
6. A pair of rudders (q) at one end of the ship is used in steering.

In the Inventor's Words

"This invention relates to a navigable balloon which is characterized essentially in that it is provided with a number of motors arranged separately from each other. In this manner it is possible to give the balloon or buoyant part of the apparatus, which receives the gas and is preferably cylindrical with rounded ends, a smaller diameter in proportion to the driving power developed by the motors and to correspondingly reduce the air resistance. A navigable balloon or air craft of this kind can be combined with several other balloons or air crafts in such a manner that the foremost craft contains the driving gear, while the others serve for the reception of the goods or loads to be carried."

The spire on top of the Empire State Building was originally planned to be used as a mooring mast for dirigibles. Gusty wind conditions at 1,350 feet rendered the plan unworkable, but it is fun to imagine looking up and seeing blimps docked over New York City.

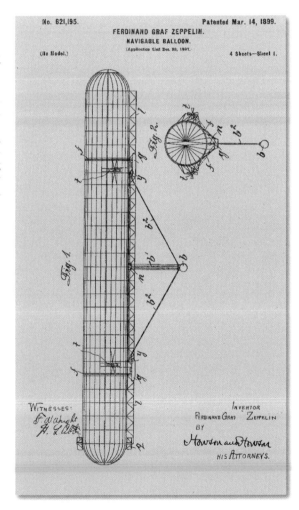

No. 621,195.

Patented Mar. 14, 1899.

FERDINAND GRAF ZEPPELIN.
NAVIGABLE BALLOON.

(No Model.)

(Application filed Dec. 29, 1897.)

4 Sheets—Sheet 1.

FIVE

The Patent That Eases

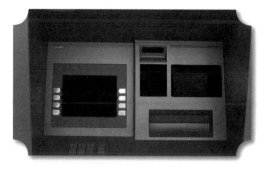

AUTOMATED TELLER MACHINE

Patent Name: Credit Card Currency Dispenser
Patent Number: 3,761,682
Patent Date: September 25, 1973
Inventors: Thomas R. Barnes, of Dallas, Texas; George R.
Chastain, of Irving, Texas; and Don C. Wetzel, of Dallas, Texas

What It Does   Dispenses cash, receives bank deposits, and relays
account information to bank customers automatically,
by means of public terminals activated by personal
magnetic cards.

Background   The idea of an ATM was not new when the first modern machine was installed as a walk-up outside a New
York bank in the early 1970s. A remote machine that
would facilitate financial transactions was the brain
child of Turkish-born American Luther Simjian back
in 1939. At that time the company that is now Citicorp
reluctantly decided to give Simjian's idea a trial run,
but the world was not ready for such mechanical convenience at the time.

"It seems the only people using the machines,"
Simjian wrote, "were a small number of prostitutes
and gamblers who didn't want to deal with tellers face
to face." Simjian would never cash in on his invention—it was removed after only six months.

A few decades later, the world had grown a little
more tech-savvy. While waiting in line at a bank, Don
Wetzel "reconceived" the ATM. Wetzel and the two

colleagues listed on his patent application worked for Docutel at the time, a company that developed automated baggage-handling equipment.

The three men developed plans for an off-line machine that would record and store information about transactions through coded information on what were the first ATM cards. But the machine was not linked by computer to the central bank. Until on-line systems were developed, banks were very selective about which clients could enjoy ATM privileges.

Today, ATMs are significantly more advanced and can perform a variety of banking functions in addition to dispensing cash. With the machines on nearly ever street corner throughout the world, stepping inside of a bank has become, for many, a thing of the past.

How It Works

1. An encoded plastic card is inserted into or "swiped through" a code-actuated automatic currency machine, which contains code-scrambling technology.
2. The user enters a personal identification code that is checked against the coded information stored in the machine.
3. Once verified, an automated dispensing apparatus reacts to user's commands.

In the Inventors' Words

"A currency dispenser automatically delivers a medium of exchange in packets in response to a coded credit card presented thereto. The coded credit card is presented to the currency dispenser and an initial check is made to determine if the card has the proper format. After checking the credit card format, coded information thereon is evaluated to check the user's identity prior to authorizing him to receive cash from the machine. When several additional checks of the credit card code have been completed, the old code is removed and substituted with a new code. The new code contains the same information as the old but updated to reflect an additional currency dispensing transaction. Both the original code and the updated code are scrambled in accordance with a changing key. Scrambling the credit card code after each use thereof minimizes the chance of unauthorized use of the currency dispenser. When the check of the credit card code indicated the user is entitled to receive the amount of currency he has selected, a storage container for the packets of currency transports the required number of packets by a positive feed drive to a cash drawer. The cash drawer opens to a detent position. . . allow the customer to then move the drawer to a fully open position to remove his currency. Upon release of the cash drawer, it returns to a partially opened position from which it automatically closes after a preset time limit."

BICYCLE

Patent Name: Improvement in Velocipedes
Patent Number: 59,915
Patent Date: November 20, 1866
Inventor: Pierre Lallement, of Paris, France

What It Does   The bicycle provides a means of personal street conveyance via a pedal-operated, wheeled vehicle for one person.

Background   For numerous generations, the bicycle has been providing sheer merriment to people of all ages. Velocipedes, as they were called in France, or bone-shakers in England, have been around for quite some time. Many individuals from a variety of countries were involved in the creation of the bike, dating as far back as the seventeenth century, long before Pierre Lallement received a U.S. patent for his in 1866. By 1895, so many patents were being awarded for variations on and modifications to the bicycle that the U.S Patent Office opened a separate department for such inventions. One popular model developed in the 1870s was the "penny farthing," which featured a high seat situated above an enormous front wheel about four times the diameter of the back wheel.

An early bicycle racer.

While there remains a great deal of dispute over the origin of the bicycle, Pierre Lallement is accepted by many as its inventor, the father of the modern bicycle. In the early 1860s, Lallement worked for a company

The first women's bicycle frame was designed by one of the Duryea brothers, the pioneers who built the first American automobile. The two siblings were originally bicycle makers. The women's frame was angled so as to allow the more petite riders to saddle more easily.

Built by Super Tandem Club Ceparana in Italy, the longest bicycle is over eighty-four feet long, and in 1998, forty club members rode it a distance of 368 feet.

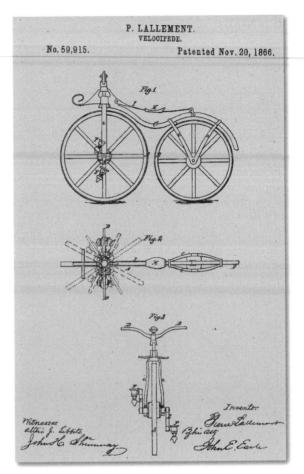

P. LALLEMENT.
VELOCIPEDE.

No. 59,915.

Patented Nov. 20, 1866.

Wheels were made of solid rubber until 1887, when John Boyd Dunlop made water-filled tires out of a garden hose, for a tricycle. He later used air, and pneumatic tires soon became the norm.

that produced wheelchairs and baby carriages, but he started toying with the idea of making his own products. Realizing others were working on similar inventions, Lallement traveled to America with some parts and gained access to a workshop in Connecticut. Here, he put the finishing touches—pedals—on what would become the first bicycle-related patent in the United States.

Today, throughout the world, bicycles provide a popular form of transportation, recreation, and sport. Appropriately enough, Lallement's home country hosts the most celebrated bike race in the world: the Tour de France.

1. A frame (C) connects the axles of two wheels (A, B).

2. The arms of the front wheel are arranged on a pivot that is controlled by handlebars (D).

3. Attached at the front axle is a pair of cranks (E) that connect to pedals (F).

4. A rider straddles a seat and begins pedaling, propelling forward while balancing his weight on the frame.

5. Braking would come a little later, but the rider did have some influence on slowing the rate of the pedal revolutions.

"My invention consists in the arrangement of two wheels, the one directly in front of the other, combined with a mechanism for driving the wheels, and an arrangement for guiding; which arrangement also enables the rider to balance himself on the two wheels."

Rothschild Petersen Patent Model Museum

This model accompanied another "velocipede" patent application, an early tricycle, awarded to E. J. Blood in 1891 (Patent Number 211,959).

BREECH-LOADING ORDNANCE

Patent Name: Improvement in Fire-Arms
Patent Number: 30,045
Patent Date: September 18, 1860
Inventor: Charles F. Brown, of Warren, Rhode Island

What It Does   Breech-loading cannons were not the most popular American Civil War–era firearms but they did introduce novel ideas in gun loading that would eventually speed up the firing action of guns. This one could be likened to an elderly ancestor of the semiautomatic machine gun.

Background   Between the years 1861 and 1865, some 10,000 sites served as settings for the American Civil War. More than 600,000 soldiers fought in it, and 2 percent of the entire American population died in its battles. In 1863, the war took a turn for the North when Union soldiers defeated Confederates at the Battle of Gettysburg, Pennsylvania. On January 31, 1865, Congress succeeded in passing the thirteenth amendment to the United States Constitution, abolishing slavery.

A breech-loading cannon in front of an armory in the 1800s.

Before and during the bloody years of the Civil War, gun makers were scrambling around trying to secure government contracts. Dozens of rifled cannons such as this one were made, and there are many classifications for them. Breech-loading (loaded through the base) and muzzle-loading were two types, but most Civil War cannon were muzzle-loading, as the breech-loaders were not yet perfected for battle. The breech-loading cannon was certainly ahead of its time. This Charles Brown invention, for example, is designed to self-load and fire as it is pushed toward enemy lines, reducing the time and labor between each explosion.

How It Works   1. A handle (H) is constructed to attach to the axles of two wheels (D) toward the front of the gun.

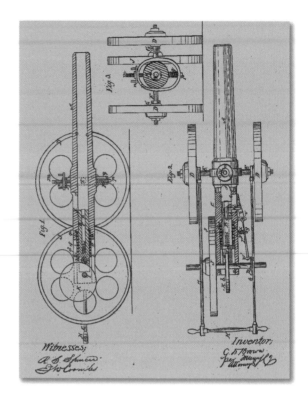

Inventor;
G. H. Brown
per Muun & Co.
Attorneys

2. A third wheel (J) is situated at back, immediately juxtaposed to the breech of the cannon.

3. As the gun is moved, the third wheel revolves its axle shaft (I), evident in figure 2.

4. The revolving shaft causes a cam (K) to drive forward a breech-pin (b) beyond apertures (f, f).

5. A number of cartridges may be arranged in a magazine above the apertures, which will drop in one by one as the breech-pin retreats and allows access to the apertures.

6. The gun can also be fired in a stationary position when the rear wheel is raised off of the ground and a crank (as shown in the model photo) is attached to one end of the shaft to rotate the wheel and move the cam.

In the Inventor's Words   "The object of my invention is to enable a piece of ordnance to be fired repeatedly by the act of its being drawn over the ground, so that it may be made very effective in advancing to meet or pursue or in retiring from before an enemy."

The original patent model.

COAL BREAKER

Patent Name: Improvement in Coal-Breakers
Patent Number: 219,773
Patent Date: September 16, 1879
Inventor: Philip Henry Sharp, of Wilkes-Barre, Pennsylvania

What It Does   The coal breaker is a large machine that is used in coal mining operations to break up the chunks of raw coal removed from the mines into pieces of a more manageable size.

Background   A finite natural resource found beneath the Earth's surface, coal is a fossil fuel that contains amorphous carbon with various organic and some inorganic compounds. In its heyday, coal fueled the Industrial Revolution in both Europe and America. It powered westward expansion on the railroad in the United States, and trade and commerce on steamships on waters throughout the world. Coal was, and still is, used to generate electricity, to heat homes, and to fire furnaces used to smelt iron ore for iron and steel.

There are four basic types of coal:

- **Anthracite** is a hard coal used for heating and generating electricity.
- **Bituminous** is a soft coal commonly used for generating electricity. As a fuel, it is not as efficient as anthracite.
- **Subbituminous** is a dull black coal found beneath the layer of bituminous coal and is used for generating electricity and heat.
- **Lignite** is another hard coal. It has a higher ash and moisture content, so its value is not as high as anthracite's.

In the mid-1800s, Pennsylvania's Wyoming Valley played a major role in the "black gold" rush. Veins of anthracite coal stretched everywhere beneath the ground in the surrounding hills; and shafts were dug left and right. Wilkes-Barre became a major railroad stop, a huge coal-mining center, and a mecca for tens of thousands of Welsh, German, English, Irish, and Russian immigrants in need of work.

These poor immigrants were exploited by the mining companies. Young boys were employed to operate coal breakers that sorted the mined coal from rock and simultaneously crushed the coal into manageable pieces. While the breaker was the bottom rung and the miner was at the top, life in the shafts was by no means glorious. Working conditions were deplorable and accidental death was an occupational hazard and a common reality.

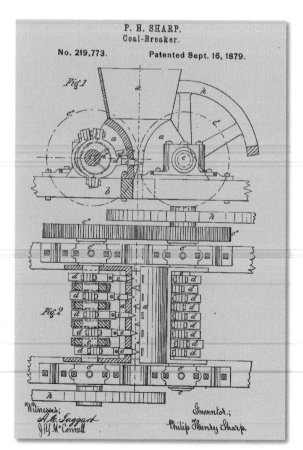

How It Works

A coal breaker in the field.

1. Set on a frame (b), two transversely secured plates (a, a) inclined toward each other feed into a hopper (x) through an opening toward their lower ends.

2. Corresponding horizontal apertures lines each of the plates, but are on separate vertical planes.

3. Parallel with the plates, two shafts (c, c) are mounted in bearings (c'); each shaft having firmly secured upon it a series of eccentrics, (d'), provided with straps (d).

4. Picks or breakers (e) are fitted to straps and secured by a key or set-screw.

5. The reciprocating picks project through the plate apertures, and act upon the coal as it is fed through the hopper.

In the Inventor's Words

Black lung disease kills more than one thousand coal mine workers annually.

"As ordinarily constructed coal-breaking machines are composed, substantially, of two parallel rollers, the peripheries of which are studded with teeth, which break up the lumps of coal fed to and passing between the rollers. Such construction is objectionable in the regard that the crushing action of the rollers pulverizes the coal to a considerable extent and entails a corresponding amount of waste, and, with a view to the avoidance of such result, the rollers have in some cases been dispensed with and the coal broken by means of vibrating plates, each carrying a series of picks or breakers operating within a stationary hopper."

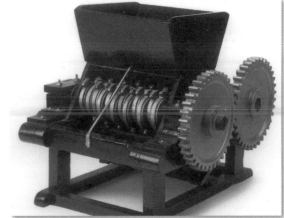

Rothschild Petersen Patent Model Museum

The original patent model.

COTTON GIN

Patent Name: Cotton Gin
Patent Number: 72X
Patent Date: March 14, 1794
Inventor: Eli Whitney Philadelphia, PA

What It Does

Mechanically cleans and separates out the seeds from cotton, so that it can be further processed for a variety of uses.

Background

One of the earliest patents in the United States was awarded to Eli Whitney for his now famous cotton gin in 1794. It is rumored that Whitney got the idea for the device when he saw a bird fly against the bars of its cage and lose feathers in the process; others claim his inspiration came from witnessing a cat trying to catch chickens in a wire coop. He envisioned that the cotton gin would streamline work and revive flagging profits for the American South's declining cotton industry— and that he might gain a share of that profit.

Along with his business partner, Phineas Miller, Whitney offered to gin the cotton for farmers throughout the South, on the condition that he got to keep a share amounting to two fifths of the profit. Instead of being heroes, though, Whitney and Miller inspired bitter feelings in the farmers, who thought their fees unfair. Farmers elected instead to make their own versions of the cotton gin and the South prospered, while little was done to protect Whitney's patent.

Whitney went north, where he had better luck making muskets. Although he didn't profit from his cotton gin, Whitney's reputation as a shrewd inventor was well established and he was able to secure a contract from the U.S. government to produce one thousand rifles. In meeting his end of the deal, Whitney honed a method for factory-producing and assembling various gun parts, much to the dismay of skilled gunsmiths and other experts in the craft.

Equality! or Yankee Ingenuity? As the availability of cotton increased, so did the monetary value of slave labor. Cotton was farmed in the South and milled and sold in the North, and resentment grew as the North began collecting taxes on the sales. Southern farmers started doing their business overseas, prompting further enmity between North and South. Some skeptical historians believe the slave controversy was no more than an outgrowth of disputes over the market of raw goods and that the moral issue was a veiled pretense by the Union to maintain an economic upper hand.

How It Works

1. Raw cotton is delivered by means of a shaft and conveyer into the main apparatus.
2. Within the apparatus, a drum cylinder rotates past a sieve, pressing against the incoming cotton.
3. The sieve separates the seed from the lint.
4. Hook-shaped claws on the drum catch the lint.
5. A rotating brush sweeps the cotton from the claws.

In the Inventor's Words

"This machine may be turned by horse or water with the greatest ease. It requires no other attendance than putting the Cotton into the hopper with a basket or fork, narrowing the hopper when necessary, and letting out the seeds after they are clean."

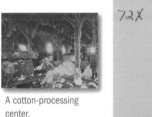

A cotton-processing center.

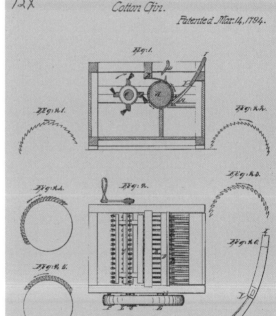

DOOR LOCK

Patent Name: Door Lock
Patent Number: 3,630
Patent Date: June 13, 1844
Inventor: Linus Yale, of Springfield, Massachusetts

What It Does

Secures a door so that entry is possible only with a key, providing security and privacy.

Background

As a young man, Linus Yale wanted to be a portrait painter, but his artistic energies soon gave way to solving mechanical problems. He decided to follow in his father's footsteps and become a lock maker. This simple door lock came before Yale achieved fame and fortune for his 1851 Infallible Bank Lock. This success allowed him to set up his own shop and continue fashioning the kind of locks that would secure banks from robbers, homes from intruders, and bathroom users from unexpected surprise entrances.

How It Works

1. A tumbler (E) is a simple cylinder of the same diameter as the cylindrical socket in which it is placed and made to revolve.

2. It is pierced in its center with a cylindrical cavity to admit a key, and perforated around its periphery with the same number of radial apertures as there are in the circular rim (C) inside the front plate of the lockbox.

3. On the revolving cylinder's perforations are placed pistons (F) that drive another set of pistons (D) back out of the apertures in the tumbler until their ends coincide with the outer periphery of the tumbler and inner periphery of the rim.

4. In this position, the tumbler can be turned for locking and unlocking the bolt.

5. A key (K) is a cylinder containing as many wedge shaped cavities or grooves (x) in its periphery as there are pistons.

6. The key also contains a cog (k) which enters a corresponding cavity or notch (n) made in the tumbler.

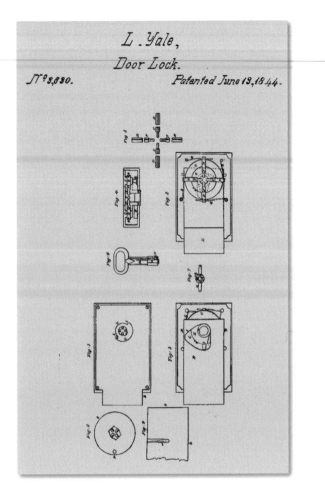

In the Inventor's Words

"The improvement consists in having a cylinder or circular rim C on the inside of the front plate of the box fastened thereto in any convenient way, or cast therewith, forming a cylindrical socket, said cylinder or circular rim being pierced from the outer to the inner circumference with round apertures on lines radiating from the center to the periphery, in which apertures are placed cylindrical pistons D, which pass through said apertures and enter corresponding apertures in a revolving cylinder or tumbler E, (having a hub or arm) that turns in said cylindrical socket for locking the bolt as hereafter described, said pistons being kept pressed inward toward the center (against other pistons F hereafter described) by springs G attached to the front plate of the box."

FLUSH TOILET

Patent Name: Detachable Flush-Rim Fixture
Patent Number: 1,107,515
Patent Date: August 18, 1914
Inventor: Philip Haas, of Dayton, Ohio

What It Does

The flush toilet is an indoor sanitation device for removing human waste. It also serves many people as a quiet place to enjoy reading material, such as the book you are now holding.

Background

The origins of the commode extend far back in history. The technology that employs running water to remove human waste is much younger. A pivotal moment in the development of the toilet came in 1775, when Englishman Alexander Cummings invented a water closet that trapped water in a U-shaped section of pipe to block the noxious odors below in what was then called a "stink trap." A valve on a hose connected to the bottom of a basin was turned to flush the debris down—and the flush toilet was in its infancy.

An early indoor flush toilet kept odors trapped inside. The next challenge was a cosmetic one.

The New World was a bit slow in adopting this technology. Several decades passed before the first American received patents for water closets. But by the end of the 1800s, great and numerous improvements would appear. This patent is among hundreds of toilet inventions developed in the United States in the first three decades of the twentieth century. It describes a detachable fixture with apertures used to manipulate water from its source and keep the bowl clean.

Eventually, toilet assemblies would include tanks and water-conserving siphons, which rely on atmospheric pressure to push water from a reservoir, over a barrier, and out the other end.

How It Works

1. Attached to a water supply pipe, the flush rim fixture is made to fit snuggly within the rear of the bowl's lip.
2. The fixture includes a series of perforations (11) that project water downward over the interior rear surface of the bowl.

3. At the fixture's extreme ends, additional orifices (12) are made to direct jets of water in opposite directions about the periphery of the bowl.

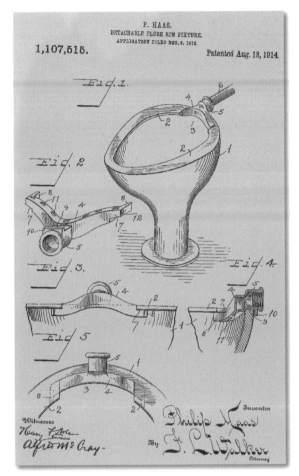

P. HAAS.
DETACHABLE FLUSH RIM FIXTURE.
APPLICATION FILED DEC. 3, 1913.

1,107,515.

Patented Aug. 18, 1914.

Thomas Crapper was an English plumber who in the mid-1800s advanced indoor toilet technology. Despite the coincidence of his name, he did not invent the contraption.

In the Inventor's Words

"Heretofore closet bowls of the type herein shown, have usually been constructed with a water conduit extending about the periphery of the bowl beneath the inturned flange thereof. This conduit is usually provided with perforations, which become clogged by deposit or sediment. Such flush conduits being located beneath the level of the bowl rim are unsanitary, since any stoppage of the soil pipe which might cause the bowl to fill or overflow, would result in the contents of the bowl filling such flush conduits and backing into the supply pipe, thus perhaps contaminating the water supply of the house."

FROZEN FOOD

Patent Name: Method of Preparing Food Products
Patent Number: 1,773,079
Patent Date: August 30, 1930
Inventor: Clarence Birdseye, of Gloucester, Massachusetts

What It Does This means of packaging and freezing fresh food insures that it remains stable over long periods and retains its flavor, texture, and color when defrosted for use.

Background While intensive research was being conducted to develop a better refrigeration system, which led to the 1928 invention of Freon, Clarence Birdseye was applying for patents related to preparing and packaging frozen foods. The scales were already tipping in his favor.

Birdsye helped usher in the era of modern convenience foods.

It makes perfect sense that Birdseye became the father of frozen foods. After quitting college where he majored in biology, he worked the U.S. Government as a naturalist in the Arctic. Birdseye saw how the native people capitalized on their environment. They kept freshly caught fish in barrels of frozen sea water, and the fish still tasted fresh when cooked long after it was caught. It froze so quickly in the subzero temperatures, he concluded, that the cellular structure of the food remained intact.

Birdseye went to New York to work on replicating this natural phenomenon with an electric fan, buckets of brine, and lots of ice. Eventually, he perfected a system of packing and flash-freezing fresh food that has inextricably linked his name to what is now a multi-billion-dollar frozen food industry.

By the end of the 1950s, frozen food sales had exceeded $1 billion; more than 60 percent of all retail grocers were stocking frozen foods; and frozen foods were being used on airline flights.

How It Works

Spinach was the first frozen vegetable to be sold.

1. Food is consolidated into a compact form.
2. The form is encased in the wrapper or carton in which it will be later sold.
3. On a system of conveyer belts, the package is further compacted by means of metal plates.
4. Simultaneously, the metal plates—maintained at temperatures of between −20° and −50° Fahrenheit—help facilitate the quick-freeze of the product as it travels through a freezing chamber.

In the Inventor's Words

"My invention relates to a method of treating food products by refrigerating the same, preferably by 'quick' freezing the product into a frozen block, in which the pristine qualities and flavors of the product are retained for a substantial period after the block has been thawed."

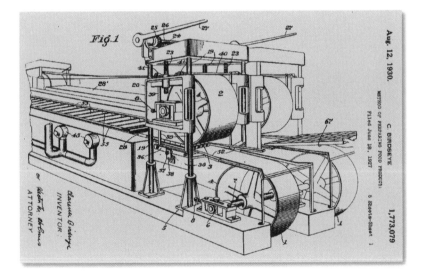

HANDCUFFS

Patent Name: Improvement in Handcuffs
Patent Number: 200,950
Patent Date: March 5, 1877
Inventor: John J. Tower and Henry W. Kahlke, of
Brooklyn, New York; assigned by John J. Tower

What It Does

Handcuffs are used primarily as a portable temporary restraining mechanism by law enforcement officers.

Background

Manacles and restraint systems have been around as long as war and crime have existed. In ancient Egypt, Jewish slaves were shackled together, often in pairs. In colonial America, petty crimes were punishable by putting the perpetrator in pillory—wooden frameworks—in the public square. For ages, people have been trying to perfect the mobile temporary restraints known as handcuffs. Around the turn of the twentieth century, the name Tower dominated the handcuff industry. John Tower took the adjustable ratchet principle from another handcuff maker of the day, W. V. Adams, who patented an earlier restraining device in 1862. This new generation of cuffs restrained criminals of all sizes. The handcuff efficiency likely annoyed apprehended criminals who remembered an earlier era of simple metal ring cuffs, more easily escapable. Tower cuffs remained the standard until the 1940s.

How It Works

1. In Figure 1, the notched segment (a) connects by joint (b) with a radius-bar (c).
2. A link (d) for the chain is also connected to the joint.
3. The hollow-bodied radius-bar contains the lock-case.
4. In Figure 2, two independent spring-catches (e, e') lock onto the notches, thereby securing the lock more firmly with and making "lock-picking" more difficult; the normal cuffs of the day had only one spring catch.
5. In figure 1, a tumbler (l) swings on a stud (3) and has notched ends.

6. A fence or stud (o) on bolt (h) prevents the key from withdrawing the bolt until the tumbler is in position for the stud to pass into the tumbler-notches.

7. In Figure 2, a cylinder (r) is grooved longitudinally to receive the plate-key and be turned. In locked position, the cylinder is held in place by a cap-plate (u).

8. The cap-plate is secured in place by studs (3, 4), and its position against the lock case itself.

In the Inventors' Words

"Handcuffs have been made with a segmental notched loop and a radius-bar containing the lock or catch. Our improvement is made to avoid the risk of the lock or catch being opened, either by an instrument introduced through the keyhole, or by a quill or thin piece of metal inserted between the segmental bar and the radius-bar."

A 1902 poster advertising "escape artist" Harry Houdini

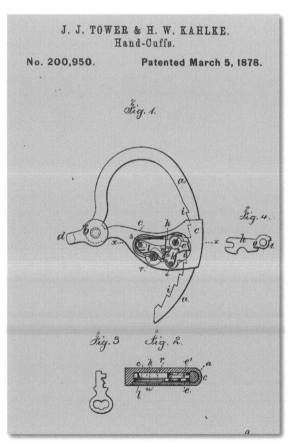

HANDLE GUIDE

Patent Name: Handle Guide
Patent Number: 2,377,745
Patent Date: June 5, 1945
Inventor: Emile Belanger, of Drummondville, Quebec, Canada

What It Does   The handle guide provides an efficient means of holding, and thereby operating, hand tools and represents
an early understanding of ergonomics.

Background   A tool should be crafted to fit as harmoniously as
possible with the hand. This is the defining principle
behind Belanger's handle guide patent. Belanger's
invention is a series of strategically placed grooves and
ridges on a tool handle meant to conform to the hand.
many computer keyboards now adhere to the same
principle, the handle guide is a close cousin to modern
day ergonomics.

Emile Belanger's invention may have been simple,
but he himself was not. He was seen in public only in
full suit and tie—no matter the season or occasion.
The rigidity of Belanger's dress code seemed tame in
comparison with his insatiable hankering for fresh seafood. According to his granddaughter, the inventor's
wife suffered a fainting spell one evening when, upon
returning home, she discovered the evening's meal
swimming in the bathtub in the form of a giant eel.

In spite of his peculiarities, Belanger was a straight-
forward thinker who enjoyed his business of building
houses in Quebec. He was quick, efficient, and had a
handle on tools.

How It Works   1. The handle (A) features a gripping portion, rounded
at the top and decreasing in diameter towards the cen-
ter.
2. Below the center is a ridge (1) extending out to
about the same diameter as the top.
3. Below the ridge is a depression which curves back
in to about the same diameter as the center of the
handle.

4. This is followed by a smaller ridge (3) which tapers out into triangular facets (4).

5. The first ridge (1) provides a firm hold for the thumb, and six cavities (5) allow for the placement of the fingers (other than the index finger).

6. The index finger, which really guides the motion of the tool, can extend further along the tool to rest firmly on any of the inner cavities (6).

Emile Belanger, inventor of the handle guide and great grandfather of this book's author.

"The operator should have proper means to guide his hand and fingers so that when he is operating the tool there should be the least variance in its motion. I have devised such a guide and in using same the operator is also saved the strain on his eyes due to the fact that he need not have his sight fixed permanently on the operation to detect or divert any faulty motion."

"It is important to note the relative alignment of these cavities for the thumb and index fingers. The division line between any two cavities (6) is opposite the centre of a cavity (5) in ridge 1. This alignment is made to correspond to the normal position the fingers should take on a handle when a tool is being properly operated."

Today, the ergonomic principles employed in this patent have become the foundations for both a science and an entire industry. The science of ergonomics considers the correlation between mechanical properties of human tissue and the response of the tissues to mechanical stress. In other words, building a more comfortable desk chair or a computer mouse that doesn't cause carpal tunnel syndrome takes advantage of the study of ergonomics.

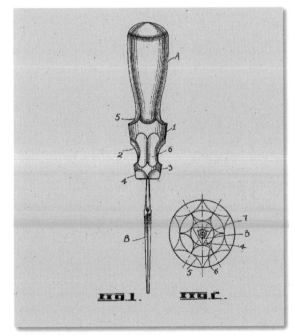

JOHN DEERE PLOW

Patent Name: Improvement in Molds for
Casting Steel Plows and Other Articles
Patent Number: 41,203
Patent Date: January 12, 1864
Inventor: John Deere, of Moline, Illinois

What It Does John Deere patented the process he used for making
the kind of plow that earned him a great reputation
in his day, and an even greater legacy in the company
that now bears his name. His secret to success was in
the way he cast molds, which resulted in an extremely
efficient plow blade.

Background The John Deere story began about forty years prior to
the date of this patent, when Deere honed his skills
as a blacksmith in Vermont. Tales of opportunities in

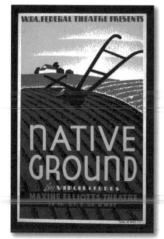

Theater poster from the '30s.

the "West"—in this case, Illinois—eventually
lured Deere to move there; but all was not
bliss. Farmers were discouraged by the way
the rich soil, in contrast to the soils back east,
clung to the plows and had to be scraped at
frequent intervals. Deere made it his mission
to solve the problem, which he did in 1837 by
using steel to craft a polished and uniquely
shaped plowshare that fastened to the mold-
board—the curved plate which cuts and
turns the earth. Until then, plows had been
made only of wood or cast iron.

But success was not instant for John
Deere. He made one plow in 1837. He made
two in 1838. He made ten in 1839. By 1840, he
still considered himself a blacksmith, according to his 1840 census form. A decade later, though,
he registered himself as "plow maker," and by the year
1858, Deere was producing one thousand plows annually. In 1864, he patented his unique idea for molding
plows. Four years and a few patents later, John Deere's
company became incorporated. Today, Deere & Company is one of the oldest industrial companies in the
United States.

A mold is formed, coated with black lead, and baked.

Deere & Company now employs about 43,000 people and does business in more than 160 countries across the globe. Its product line has expanded considerably to include everything from riding mowers and skid steer loaders to forestry excavators and dump trucks. In 2002, Deere & Company's annual sales and revenues were $13.9 billion. The company marked the 200th birthday of company founder John Deere on February 7, 2004.

"I form a mold by means of a suitable pattern out of a dry sand in the usual way, and after the mold is formed I coat its inner surface with black-lead (plumbago) which may be moistened by water having a little fire-clay dissolved in it. The sand also used in forming the mold may be tempered with fire-clay water. This fire-clay water has a tendency to render the sand and black-lead or plumbago rather more adhesive than they would be if water alone were used. After the mold is formed and coated internally with the black-lead or plumbago, it is baked, say, for about twelve hours, or until all moisture is thoroughly expelled from it, and when thus baked and thoroughly dried the black-lead or plumbago will form a hard crust, coat, or glazing for the inner surface of the mold, so as to prevent the melted steel from coming in contact with the sand. The latter therefore cannot be fused and cut away by the melted steel, and will form or produce a perfect casting, or one with a smooth surface and free from holes or air-cells. This mold may, before being baked and dried, be perforated with vent-holes to admit of the escape of air and gases. I would further remark that the mold is used but for one casting only, it being, after once used, broken up or pulverized and the same ingredients used for the formation of another mold."

LAWN MOWER

Patent Name: Lawn Mower
Patent Number: 624,749
Patent Date: May 9, 1899
Inventor: John Albert Burr, of Agawam, Massachusetts

What It Does This simple machine keeps a lawn trimmed uniformly to the desired height with a minimum of physical labor, by means of blades that rotate as the device is propelled forward.

The rotary blade lawn mower was heavy and hard to push up hill, but at least it was quiet.

Edwin Budding of Gloucestershire, England, patented the first "machine for mowing lawns" in 1830. Budding's design was based on carpet-trimming tools, with multiple blades attached side-by-side around a shaft.

Connecticut resident Amariah Hills received the first U.S. patent for a reel lawn mower in 1868; the growing expanse of suburbs along the East Coast of the United States was creating demand for an efficient machine of this sort.

The first rotary blade lawn mower was created by fellow Connecticut resident John Albert Burr in 1899. The contribution of African-American inventor John Albert Burr to the device was some significant components that set a new standard for a safer and better lawn mower. He included a set of three curved rotary shears that prevented the freshly clipped grass from clogging the gears, and traction wheels set within the reach of a longer blade. This design allowed the mower to access to the grass while reducing the overlap caused by previous wheel positions.

How It Works

1. Wide traction wheels (A) are set at the inside ends of an axle shaft (B).

2. Cheek-plates (C) secured on the outer edge, the hub, of each tire support an outward-extending central shaft (f), the fixed bar of the cutter (k), and a countershaft (i).

3. Loosely mounted on the axle shaft is a gear-wheel (O), which engages a ratchet wheel that synchronizes the rotation of the blade with the motion of the traction wheels.

4. As the operator pushes the device forward, the blade rotates and shears the grass in a uniform swath.

In the Inventor's Words

"By making the side cheek-plates with the forward extension members decidedly out-forwardly offset beyond the outer faces of the traction-wheels provides for the mounting between the forward ends of the cheeks of a rotary cutter having a length quite a little exceeding the distance between the outer faces of the traction wheels, and hence it becomes possible to cut the grass more closely to a building or other object than would be possible were the ends of the rotary cutter disposed within the planes coincident with the outer faces of the said wheels, and it will be seen that in the use of the machine its every pass will cut the grass in a path wider than that comprised between the traction-wheels, so that the disadvantage of having one of the wheels bed down the grass just outside of the cut path or swath at every pass is overcome."

An alternative to more costly motor sports, the first British Grand Prix Lawn Mower Races were held in 1973, with separate races for run-behind mowers, towed seat mowers, and tractor seat mowers. Seven years later, Tony Bell's name was added to the Guinness Book of World Records for covering the greatest distance in a single lawn mower race—276 miles.

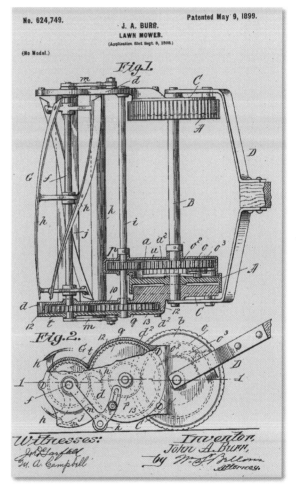

No. 624,749.

Patented May 9, 1899.

J. A. BURR.
LAWN MOWER.
(Application filed Sept. 8, 1898.)

(No Model.)

MICROWAVE

Patent Name: Method of Treating Foodstuffs
Patent Number: 2,495,429
Patent Date: January 24, 1950
Inventor: Percy L. Spencer, of West Newton, Massachusetts

What It Does

Cooks food quickly by microwave radiation.

Background

Many urban legends followed the introduction of the microwave. For example: An elderly lady who bred Persian cats used to wash her cat, towel it dry, and then warm it in her electric oven. Her son brought her a brand-new microwave oven for Christmas. On the day of the next cat show, not understanding the technology, she washed her prize-winning Persian cat and popped it into the microwave for a few seconds. There was no time for a meow. The cat exploded the instant the oven was switched on. Poodles, babies, and exploding glasses of boiling water have all had their turn in the urban folk tale microwave.

In 1946, Percy Spencer, a self-taught electronics wizard who never graduated grammar school, received U.S. patent 2,408,235 for the development of a high efficiency magnetron. Magnetron tubes employ heated cathodes to generate electrons that are affected by electro-magnetic energy and produce microwave radiation. Microwave radiation is used both in radar and in the "method of treating foodstuffs" that Spencer invented and patented four years later.

While working for the Raytheon Manufacturing Company, Spencer was running tests on one of the new magnetron tubes when he noticed the candy bar in his pocket begin to melt. Employing his natural inventor instincts, he decided to place some uncooked popcorn in front of the magnetron tube. To his delight, the kernels began popping. (This is still a favorite application of microwave technology.)

Engineers at Raytheon went to work on Spencer's discovery. The first commercial microwave oven was introduced in 1947. It was 750 pounds and five-and-a-half feet tall. Twenty years and many modifications later, the first countertop, domestic microwave oven hit the market. By 1975, microwaves were selling faster than conventional gas-range ovens.

How It Works

1. Electron-discharging cathodes (14) housed within oscillators are joined by conductors (20, 21).
2. These conductors are in turn connected by another conductor (22), a center tap on a winding (17) of the transformer (18).
3. Coaxial transmission lines (24, 25) alternately deliver hyper-frequency energy to a wave guide (23).

4. The wave guide is directed at a piece of food on a speed-adjustable conveyor (28).

"Such energy has been used before for this purpose, but the frequencies employed have been relatively low, for example, not over 50 mega-cycles. I have found that at frequencies of this order of magnitude, the energy necessarily expended in order to generate sufficient heat to satisfactorily cook the foodstuff is much too high to permit the practical use of the process. I have further found, however, that this disadvantage may be eliminated by employing wave lengths falling in the microwave region of the electromagnetic spectrum, for example, wave lengths of the order of 10 centimeters or less. By so doing, the wave length of the energy becomes comparable to the average dimension of the foodstuff to be cooked, and as a result, the heat generated in the foodstuff becomes intense, the energy expended becomes a minimum, and the entire process becomes efficient and commercially feasible."

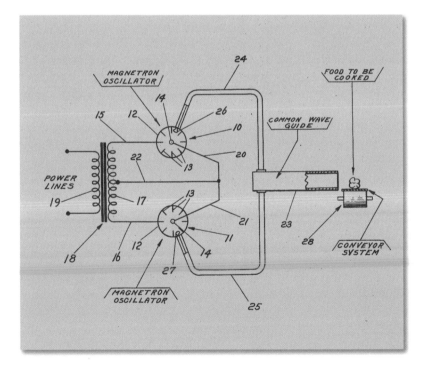

PASTEURIZATION

Patent Name: Improvement in Brewing Beer and Ale
Patent Number: 135,245
Patent Date: January 28, 1873
Inventor: Louis Pasteur, of Paris, France

What It Does
This heat-based process, named for its inventor, kills bacteria in perishable liquids such as beer and milk, and stabilizes and preserves them without significantly compromising taste or quality.

Background
Louis Pasteur was a brilliant scientist whose work is memorialized by the semisterilization process named in his honor. The French scientist disproved the long-held theory of spontaneous generation, which maintained that microorganisms arise in decaying biological matter. Pasteur believed that microorganisms existed everywhere; that they were not symptoms of decay but rather the cause, and that the contamination of food and beverages by bacterial microbes could be prevented through a specialized heat treatment that minimized exposure to microbe-carrying air, yet did not destroy the food.

Louis Pasteur

Pasteur put his theories into practice by coming up with an improved process of fermentation. Previously, fermentation of sugar by yeast had yielded the desired alcoholic product, but also some undesirable by-products. Pasteur believed these were related to particular microbes, or "germs," which were responsible for fermentation, but could be separated out and treated for a better result. He also demonstrated that particular germs were responsible for many types of diseases. In their weakened forms, some microbes could be used for immunization purposes.

Three of the inventor's five children died from typhoid fever, and Pasteur was determined to unlock the mysterious causes of disease. He was quite successful, too. Pasteur found that rabies was transmitted by viruses that had previously been undetected because they were invisible even under a microscope. He subsequently developed a vaccination against rabies. Altogether, Pasteur's work revolutionized food sterilization and disease treatment, and he single-handedly initiated various new fields of medical and scientific study.

Why does draft beer taste fresher? Because most commercial breweries bypass pasteurization of keg beer (which can reduce flavor) and insist on keeping it properly refrigerated.

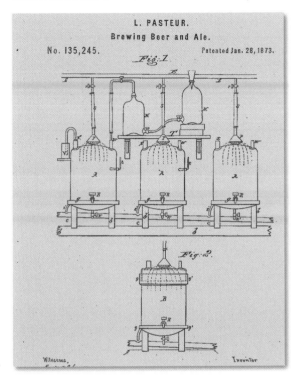

How It Works

Pasteurization plant

Today's large commercial breweries still use pasteurization. Beer is aged in tanks, filtered, artificially carbonated, and then pasteurized, which preserves the beer's freshness by eliminating spoilage organisms or residual brewing yeast that can compromise the quality of the beer. The process Pasteur developed also opened the doors for bottling, allowing for at home enjoyment of the beverage.

1. In the pasteurization of beer, boiling hot wort—unfermented malt infusion—is introduced into a tank (A).

2. From gas supply tanks (M), carbonic-acid gas is projected into the tank through piping (w) to remove air from the wort.

3. A water supply pipe (E) delivers spray through nozzles (P) to the outside of the tanks to cool the wort.

4. The wort is left or delivered into another vessel, where it ferments.

"My invention has for its object to produce a better quality and greater quantity of beer from the same quantity and quality of wort, and to afford a beer which shall also embody the quality of greater degree of unalterableness during time and changes of climate &c., in transportation and use; and to these ends my invention consists in expelling the air from the boiled wort while confined in a closed vessel or closed vessels, and then cooling it by the application of sprays of water to the exterior of such vessel or vessels."

POLYGRAPH LIE DETECTOR

Patent Name: Apparatus for Recording Arterial Blood Pressure
Patent Number: 1,788,434
Patent Date: January 13, 1931
Inventor: Leonard Keeler, of Berkeley, California

What It Does

The polygraph lie detector measures a variety of physiological fluctuations while an individual is undergoing questioning for investigative purposes, in order to help determine whether the person is answering the questions truthfully.

Background

Administering a lie detector test

The art of lie detection was introduced to science in 1921. A University of California medical student named John Larson invented the first polygraph lie detector—a machine that measured blood pressure, pulse, and respiration changes in response to questioning. The basis of the device is the belief that humans, when lying while under investigation, produce revealing involuntary physiological reactions that can be detected by experts. Leonard Keeler was a criminologist who ardently supported Larsen's invention and the theory behind it. He subsequently refined the invention, patented it, and marketed it commercially. Keeler's invention—initially used for medical and psychological testing only—was a primitive yet significant first step in the evolution of the polygraph.

While polygraph lie detectors are commonly used to interrogate crime suspects, they are not considered completely reliable. Some argue that the most criminal-minded individuals are free from relatively normal psychological constraints and are thereby able to control their physiological responses in order to "beat" the machine. On the other hand, individuals telling the truth under pressure of interrogation might show physiological signs that incorrectly implicate them. Lie detection remains an art, and the polygraph lie detector test is not generally admissible as evidence in court.

How It Works

1. A sphygmograph attaches to a person's body to measure blood pressure.
2. It is also connected to a recording device.
3. Fluctuations in pressure greater than those caused by the normal cardiac cycle are registered on paper by the recording device.

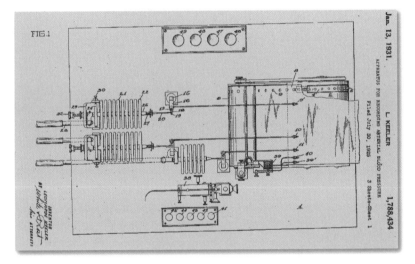

In the Inventor's Words

"It is an object of the invention to provide means whereby the sphygmogram or cardiac cycle may be recorded simultaneously with and be superimposed on the slower oscillations in the arterial pressure, whereby the characteristics of each as well as their relation to each other at any moment may be ascertained."

POST-IT NOTES

Patent Name: Acrylate Microsphere-Surfaced Sheet Material
Patent Number: 3,857,731
Patent Date: December 31, 1974
Inventors: Roger F. Merrill, Jr., of Troy Township, Wisconsin,
and Henry R. Courtney, of St. Paul, Minnesota; assigned
by Minnesota Mining and Manufacturing Company

What It Does   The Post-It note is one of the most quickly adopted and widely used office innovations of all time. Coming in different colors and sizes, the notes adhere to documents, computer monitors, doors, windows, and desk drawers throughout the world to help individuals on every rung of the corporate ladder. They are used to remind interns to fax documents, CEOs to attend board meetings, and husbands to buy milk.

Background   In 1970, research was being conducted at the Minnesota Mining and Manufacturing Company labs to improve acrylate adhesives that the company used in many of its tapes. It was this company, 3M, after all,

which first developed cellophane tape. Spencer Silver came up with an adhesive that didn't quite work—his stickum in the form of tiny spheres was just too weak to adhere well to any tape backing because of the intermittent nature of the contact the spheres made with the backing. Despite the fact that his invention was considered too feeble to have any commercial value, Silver didn't give up on it. He lectured on its possibilities and praised its potential as a spray adhesive. His "acrylate copolymer microspheres" invention was awarded US Patent #3,691,140 in 1972.

Shortly thereafter, Art Fry, one of 3M's product development researchers, came to see the value in Silver's adhesive. It could be used to make a better bookmark. Corporate folklore holds that Fry was always frustrated by losing his place in the hymnal at church. From Fry's ideas others flowed, and soon The Post-It note was born, hitting office-supply store shelves in 1980.

How It Works

1. Elastomeric copolymer spheres (30), range in diameter from 1 to 250 microns.

2. These are set upon a substrate (20) which adheres to a bind (30).

3. The substrate is not so porous as to allow the elastomeric copolymer spheres to permeate its surface completely.

4. Some substrates listed in the patent include films such as polyesters, cellulose acetate, and polyvinyl chloride, glass, wood, vinyl copolymers and urethane cast closed cell foams, and paper.

Fashion designer Ilze Vitolina created a line of avant-garde evening wear fashioned from Post-it notes. To create the clothing, the notes were placed on a durable plastic, and plastic strips were used as fasteners. Vitolina created eleven dresses—including a wedding dress—as well as several hats and a bridal bouquet. The dresses were modeled at a 2000 fashion show sponsored by 3M Latvia.

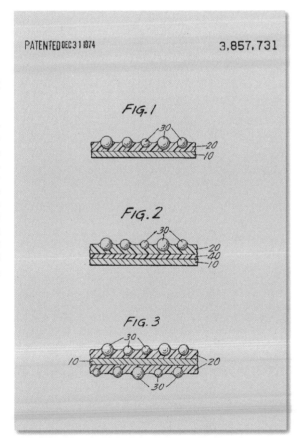

PATENTED DEC 31 1974   3,857,731

FIG. 1

FIG. 2

FIG. 3

In the Inventors'
Words

"The invention relates to pressure-sensitive sheet constructions. More particularly, it relates to sheet constructions containing an adhesive system which permits repeated cycles in which materials are bonded thereto and removed therefrom.

"Conventional adhesives for adhering paper and other like materials to substrates, while having many desirable features, also have inherent drawbacks. For example, while some such adhesives may permit removal of paper from a substrate to which it has been adhered, they do not permit rebonding of the paper to the substrate. Conversely, other adhesives possess a tack which may be so aggressive as to cause the paper to tear on removal."

REMOTE CONTROL

Patent Name: Ultrasonic Transmitter
Patent Numbers: 2,821,954 and 2,821,955
Patent Date: February 4, 1958
Inventors: Robert Adler, of Northfield, Illinois; Robert C. Ehlers, Lombard Villa, and Clarence W. Wandrey, of Wheaton, Illinois; assigned by Zenith Radio Corporation

What It Does

These two patents represent the innovation introduced as the "Zenith Space Command," in 1956—the first commercially successful wireless remote control device for controlling a television signal from a short distance.

Background

Prior to the "Zenith Space Command," the company made the "Lazy Bones" remote, which was attached by a cable to the TV set. Then, Zenith introduced the "Flashmatic"—the first wireless remote, which used photocells and had many drawbacks. Mildly stated, the "Flashmatic" was erratic. Austrian-born Zenith research director Robert Adler championed the development of ultrasonic technology to perfect the remote. Ultrasonic waves—sounds that extend beyond the range of human hearing—provided the standard for two-

and-a-half decades of channel surfing until infrared light technology was introduced in the early 1980s.

Manipulating one's televised entertainment was not so easy before the remote control. Today, many people in privileged societies cannot imagine having to actually get up to change the channel, and legions of couch potatoes are probably unaware of the debt they owe Robert Adler. While Adler was awarded more than one hundred patents for a variety of technological innovations, he is regarded in the television industry as the father of the remote control.

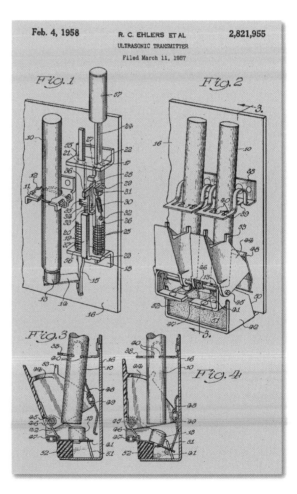

Feb. 4, 1958   R. C. EHLERS ET AL   2,821,955
ULTRASONIC TRANSMITTER
Filed March 11, 1957

How It Works Before transistors began replace vacuum tubes, the tubes in the remote-activated TV sets were designed to receive and process the ultrasonic signals transmitted by the remote control.

1. The transmitter consists of four aluminum rods.

2. Each rod is carefully cut to lengths to generate a distinctive ultrasonic frequency.

3. Buttons on the remote control unit engage a tiny spring-activated hammer that strikes the corresponding aluminum rods, thereby creating the high-frequency sounds.

4. Each of the rods has a distinct function: one turns the channel up; one turns it down; one increases volume; one decreases volume.

TAMPON

Patent Name: Catamenial Device
Patent Number: 1,926,900
Patent Date: September 12, 1933
Inventor: Earle C. Haas, of Denver, Colorado

What It Does Provides handy, hygienic, and undetectable internal means of absorbing menstrual flow.

Background Some form of tampon has been in use for ages. While the men of ancient Egypt were covering the walls of pyramids and obelisks with hieroglyphics, their women were crafty enough to insert plugs of softened papyrus to absorb their menses. Women all over the world have used whatever natural material was handy for this purpose. External cloth pads eventually became the prevailing preference, and these "sanitary napkins" became commercially available in the 1920s—but they were neither practical nor comfortable.

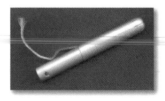

At the same time, doctors regularly used makeshift plugs of cotton to apply antiseptics in bodily cavities, to stanch hemorrhaging, and to absorb blood during surgical procedures. It was in fact an observant physician, Dr. Earle Cleveland Haas, who put two and two together—his sympathy for his wife's dissatisfaction with her pads led him to invent the first disposable tampon with an applicator, very similar to those used today. Calling upon the Greek term for the monthly menstrual discharge, he called his invention a "catamenial" device, but the parlance changed when he registered a trademark for his invention. Combining the words "tampon" and "vaginal pack," Haas put Tampax on the market, where it has remained ever since.

Toxic shock syndrome came to light in 1980, after the advent of a highly absorbent new tampon product called Rely. TSS is caused by the common bacteria—*Staphylococcus aureus*. Certain strains of this bacteria can produce toxins that poison the blood, causing nausea, fever, diarrhea, mental confusion, rash, and in some cases, death. It is believed that leaving a tampon in too long or using one with higher absorbency levels may allow the rare strain of bacteria to flourish. Rely was removed from shelves in 1981, but warnings are now printed on the packaging of all tampon products, advising frequent changing and the monitoring of possible symptoms. Dozens of women have died from toxic shock syndrome, but most women are willing to assume the risk in order to avail themselves of this comfortable and convenient product.

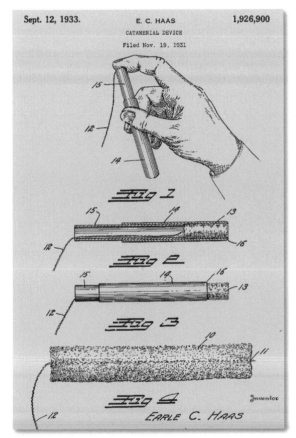

Sept. 12, 1933.   E. C. HAAS   1,926,900

CATAMENIAL DEVICE

Filed Nov. 19, 1931

Inventor

EARLE C. HAAS

How It Works

1. A seam (11) is sewn through an elongated strip of cotton (10).

2. An additional length of string (12) extends beyond the strip.

3. The strip is compressed under high pressure to form and retain a cylindrical core (13) until use.

4. The core is placed in an outer tube (14).

5. Another tube (15) is placed within the outer tube, protecting the sterile core.

6. A rounded edge (16) on the outer tube's edge increases comfort upon insertion.

7. The inner tube operates as a plunger, facilitating insertion, after which both tubes are removed and disposed of.

8. Upon removal, the string (12) provides a means of drawing the tampon out.

In the Inventor's Words

"The principal object of this invention is to provide an absorbent pad and means for conveniently inserting said pad into the female vagina.

"Another object of the invention is to combine as a single unit an absorbent pad for insertion in the vagina with an inserting applicator, which acts as a container for the pad, so that the pad can be furnished in the applicator and inserted without removing it therefrom and so that the applicator can be discarded after use."

TRAFFIC SIGNAL

Patent Name: Traffic Signal
Patent Number: 1,475,024
Patent Date: November 20, 1923
Inventor: Garrett A. Morgan, of Cleveland, Ohio

What It Does

Signals hung prominently at active intersections provide an orderly and safe traffic flow.

Background

Shortly after automobiles began rolling out of factories and onto public streets, traffic safety became a major issue. Until then, horse-drawn carriages and bicycles shared the roadway in relative harmony. A collision between an automobile and a carriage was the impetus for an African American named Garrett Augustus Morgan to invent the traffic signal.

The son of former slaves, Morgan was born in Kentucky and moved to Cincinnati and then to Cleveland while still a teenager. He worked as a sewing machine repairman, developing a reputation as a crafty and innovative mechanic. He ultimately became a successful businessman with a notable concern for public safety.

Though it wasn't the green, red, and yellow signal now used universally, Morgan's traffic signal did consist of three separate indicators: stop, go, and an all-direction stop to allow for street-crossing pedestrians. This last indicator also served the purpose of the current yellow light: a means to warn drivers that the flow of traffic was changing. A traffic conductor easily operated the signal by turning a crank.

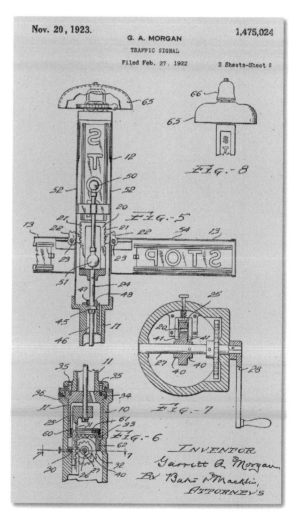

Garrett Morgan also invented a gas mask that proved sensationally effective on July 15, 1916, when he used it to rescue people in an underground tunnel beneath Lake Erie after an explosion occurred.

How It Works

1. A rigid, boxlike structure is mounted atop a supporting pole at an active intersection.
2. Indicators feature the words "Stop," "Go," and "All Stop" on alternating panels.
3. The panels are supported by pivots operated by a hand crank.
4. The bottom of the supporting pole is affixed into a stand, out of which the crank handle protrudes. An operator uses the crank to raise the indicating arms, by means of ratchets; to lower them; and to revolve them.
5. The direction-indicating arms are pivotally supported and adapted to move vertically for stopping the flow of traffic in one direction; and then revolved and dropped to indicate a right of way to vehicles moving in another direction.
6. The box on the top is rotated by turning the crank, which engages a circular ratchet mechanism, so that the panels with the traffic instructions can face alternating different directions.
7. Traffic is thus directed by raising the arms to stop traffic in one direction, whereby all traffic is stopped; then revolving the arms, which also rotates the top box, and releasing them to indicate right of way for traffic in another direction.

In the Inventor's Words

"One of the objects of my invention is the provision of a visible indicator which is useful in stopping traffic in all directions before the signal to proceed in any one direction is given. This is advantageous in that vehicles which are partly across the intersecting streets are given time to pass the vehicles which are waiting to travel in a transverse direction; thus avoiding accidents which frequently occur by reason of the over-anxiety of the waiting drivers to start as soon as the signal to proceed is given."

The inventive contributions of African-Americans are numerous and inspiring. Here is a list of just a few important ones:

• Thomas L. Jennings patented a process of dry-scouring clothing. With his earnings, he was able to extricate himself and his family from the bonds of slavery.

- A nuclear physicist, George E. Alcorn, invented an imaging X-ray spectrometer.

- After quitting school at age ten, Granville T. Woods invented the incubator and made a career move from firefighter to engineer.

- Hermon Grimes invented the folding wing aircraft, the design of which was used in a craft flown by George Bush Sr. during World War II.

- Richard B. Spikes invented the automatic gear shift and a braking device still used in school buses.

- Dr. Charles Richard Drew was the first African-American surgeon to serve as an examiner on the American Board of Surgery. He invented the concept of blood banks.

- Sarah E. Goode was the first African-American woman to receive a patent; she invented a folding cabinet bed.

VACUUM CLEANER

Patent Name: Carpet Sweeper and Cleaner
Patent Number: 889,823
Patent Date: June 2, 1908
Inventor: James M. Spangler, of Canton, Ohio

What It Does   Sucking dirt, dust, and debris out of the carpets and furnishings of homes, hotels, and offices around the globe, the vacuum cleaner is one of the handiest sanitation mechanisms ever invented.

Background   An asthmatic department store janitor named James Spangler invented the modern electric vacuum cleaner in 1907, and established the Electric Suction Sweeper Company. The president of the company was an enthusiastic customer of Spangler's product and also

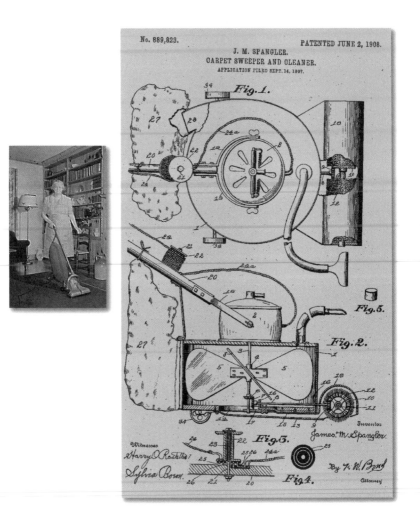

his cousin's husband—a man named William Hoover. Hoover is credited with the concept of door-to-door salesmanship. To attract attention to the product, Hoover offered in-home demonstrations. In 1922, he changed the company name to "Hoover." Spangler's name, on the other hand, got sucked out of the spotlight.

How It Works

1. An electric motor (2) attaches to a fan casing (1).
2. From the motor, a shaft (3) protrudes into the casing, where it attaches to a hub (3).
3. The hub is securely connected to the fan blades (5), which are arranged, when rotating, to create suction at the casing's bottom.

4. The casing sits on back wheels (34) and a gearwheel cylinder covered in a circular brush (12).

5. Also at the bottom, a plate (13) is arranged to funnel debris into the machine through an inlet opening (14).

6. Dust and debris are then blown through an outlet (28, in figure 1) and into a collecting sack (27).

In the Inventor's Words

"In a sweeper, the combination with a casing formed with a centrally disposed opening in its bottom, a bearing extending over the opening, a vertical shaft journaled into the bearing, a fan mounted on the vertical shaft, a flat housing which tapers toward the front and located under the casing which communicates with the centrally disposed opening, said housing extending beyond the casing and formed with an elongated opening in its bottom at the widest portion, a brush mounted in the extended end of the housing adjacent the elongated opening, means of communicating with the casing for receiving the dust laden air, and means for revolving the brush."

VIAGRA

Patent Name: Pyrazolopyrimidinodes which Inhibit Type 5 Cyclic Guanosine 3',5' –Monophosphate Phosphodiesterase (cGMP PDE5) for the Treatment of Sexual Dysfunction.
Patent Number: World Intellectual Property Organization 98/49166-A1
Patent Date: November 5, 1998
Inventors: Mark Edward Bunnage of Sandwich, Kent, United Kingdom; John Paul Mathias, of Sandwich, Kent, United Kingdom: Stephen Derek Albert Street, of Sandwich, Kent, United Kingdom; and Anthony Wood, of Sandwich, Kent, United Kingdom: All patentees assigned by Pfizer Central Research.

What It Does   Viagra and other similar drugs are becoming an increasingly popular treatment for erectile dysfunction. The drug essentially enables a fuller flow of blood to the penis, promoting more reliable and longer-lasting erections.

Background   When Masters and Johnson released their revolutionary studies of sexuality in the 1950s, private topics suddenly became the subject of public discourse. This helped open wide the door to medical studies that focused on sexual behavior in general and reproductive organs specifically. Not surprisingly, commercial markets sprang up around products that held the promise of enhanced sexual satisfaction.

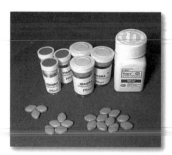

Female Sexual Dysfunction (FSD) is another well-documented problem, and Viagra has been helping women, too. New products are under development to address specifically female problems.

The brainchild of researchers at the pharmaceutical giant Pfizer, Viagra is one such product for which the company has fought long and hard. A medicine used to treat erectile dysfunction (ED), Viagra first became available in 1998. Pfizer boasts that its product "has helped more than 16 million men around the world improve their sex lives. Amazingly, 9 tablets are dispensed every second worldwide."

Clearly, Viagra is a money maker and the manufacturer has fought to monopolize the market for it. The United Kingdom's High Court ruled that Pfizer's European patent (involving the use of sildenafil, Viagra's

Wrigley's Gum was recently granted a U.S. patent for a gum that delivers sildenafil citrate, though they have not announced plans on marketing it. Who knows? ED sufferers may soon be chewing their way to better sex lives.

active ingredient) was invalid. The legal challenge was brought by Eli Lilly and ICOS Corp, who were co-developing Cialis, a rival impotency drug. Other rival medications are in the works, and some are on the shelves already. The battle lines have been drawn and the swords are raised. In October 2003, Pfizer filed a lawsuit against GlaxoSmithKline and Bayer Pharmaceuticals Corporation for patent infringement, for the marketing of Levitra.

How It Works

Sildenafil, the active ingredient in Viagra, acts like a traffic director who shrewdly opens lanes in one direction during rush hour. The compound enables arteries in the penis to relax and widen to offer more access for blood flow towards the organ. Simultaneously, the drug restricts the veins that carry blood away from the penis. Altogether, sildenafil is known to achieve solid results during sexual intimacy.

In the Inventor's Words

Blue Versus Yellow
In 2003, the NFL became the first professional sports league to sign an endorsement deal allowing GlaxoSmithKline and Bayer Pharmaceuticals Corporation to market a Viagra rival called Levitra. Levitra, which comes in the form of a little yellow pill, may arouse some stiff competition against Pfizer's little blue pill. Many touchdowns are anticipated.

"Compounds of formulae (IA) and (IB) or pharmaceutically or veterinarily acceptable salts thereof, or pharmaceutically or veterinarily acceptable solvates of either entity, wherein R_1 is C_1 to C_3 alkyl substituted with C_3 to C_6 cycloalkyl, $CONR_5R_6$ or a N-linked heterocyclic group; $(CH_2)nHet$ or $(CH_2)nAr$; R_2 is C_1 to C_6 alkyl; R_3 is C_1 to C_6 alkyl optionally substituted with C_1 to C_4 alkoxy; R_4 is $SO_2NR_7R_8$; R_5 and R_6 are each independently selected from H and C_1 to C_4 alkyl optionally substituted with C_1 to C_4 alkoxy, or, together with the nitrogen atom to which they are attached, form a 5- or 6-membered heterocyclic group; R_7 and R_8, together with the nitrogen atom to which they are attached, form a 4-R_{10}-piperazinyl group; R_{10} is H or C_1 to C_4 alkyl optionally substituted with OH, C_1 to C_4 alkoxy or $CONH_2$; Het is an optionally substituted C-linked 5- or 6-membered heterocyclic group; Ar is optionally substituted phenyl; and n is 0 or 1; are potent and selective cGMP PDE5 inhibitors useful in the treatment of, inter alia, male erectile dysfunction and female sexual dysfunction."

To all whom it may concern:

Be it known that I, FRANZ VESTER, of Newark, in the county of Essex, and State of New Jersey, have invented a new and useful Improvement in Burial-Cases or Coffins for the dead; and do hereby declare that the following is a full, clear, and exact description of the same, reference being had to the accompanying drawings, making part of this specification, of which—

Figure 1 is a top view; Fig. 2, a side elevation; Fig. 3, an under side view of the lid; and Fig. 4, a longitudinal section.

The nature of this invention consists in placing on the lid of the coffin, and directly over the face of the body laid therein, a square tube, which extends from the coffin up through and over the surface of the grave, said tube containing a ladder and a cord, one end of said cord being placed in the hand of the person laid in the coffin, and the other end of said cord being attached to a bell on the top of the square tube, so that, should a person be interred ere life is extinct, he can, on recovery to consciousness, ascend from the grave and the coffin by the ladder; or, if not able to ascend by said ladder, ring the bell, thereby giving an alarm, and thus save himself from premature burial and death; and if, on inspection, life is extinct, the tube is withdrawn, the sliding door closed, and the tube used for a similar purpose.

In the said drawings, A denotes the body of the coffin; B, the lid. C represents a square tube, which is seated in a square base, D, attached to the lid of the coffin, and held in place by a spring-bar, E, connected with the sliding glass door L. This square tube C extends from the lid of the coffin to and above the surface of the grave, and has air-inlet openings F F, which communicate with the body of the coffin, and has also a glass door, K, on its top, which may be easily raised or looked through, for inspection of the person laid in the coffin. The said tube contains a

ladder, H, by which the person laid in the coffin may, on returning life, ascend to the surface of the earth; and the said tube C has near its top a bell, I, from which a cord is suspended, the lower end of said cord being placed in the hand of the person laid in the coffin, as shown in the drawings. On the face of the coffin-lid is a sliding glass door, actuated by a spring, M, which closes the coffin, excluding the air when the tube C is drawn from the coffin.

The operation of my invention is as follows: The corpse being laid in the coffin A of the coffin, and the cord K placed in the hand of the corpse, the cord is next passed through the tube C and attached to the bell I, and the tube C placed in the base D on the lid of the coffin. The coffin is now let down into the grave, and the grave filled up to the air-inlets F F. Now, should the person in the coffin, on returning life, desire to ascend from the coffin and the grave to the surface, he can do so by means of the ladder, but, if too weak to ascend by the ladder, can pull the cord in his hand, and ring the bell I, giving the desired alarm for help, and thus save himself from premature death by being buried alive. Should life be extinct, the tube C is removed, the door L closed, and the tube used for a similar purpose.

Having described my invention, what I claim as new, and desire to secure by Letters Patent, is—

1. The application of the tube C and ladder H to a burial-case or coffin, substantially and for the purposes described and set forth.

2. In combination with the tube C and ladder H, the cord K and bell I, for the purposes substantially as set forth and described.

In testimony whereof I have hereunto set my signature this 9th day of July, 1868.

FRANZ VESTER

Witnesses:
    A. NEILL,
    R. SANGMEISTER.

SIX

The Patent That Pleases

James Naismith

BASKETBALL

Patent Name: Basketball
Patent Number: 1,718,305
Patent Date: June 25, 1929
Inventor: George L. Pierce, of Brooklyn, New York

What It Does   This sturdy, well-designed, uniformly manufactured, and long-lasting basketball is the prototypical center-piece to the modern game of basketball, both professional and amateur.

Background   The modern game of basketball was invented by Canadian James Naismith, a graduate of McGill University in Montreal. He took a job as a physical education instructor at the School for Christian Workers in Springfield, Massachusetts. A strict boss gave him orders and a short deadline to come up with a way to occupy a class of rowdy kids during the fast-approaching winter of 1891–1892. Naismith came up with a set of rules and a game that is the closest known precursor to the modern sport of basketball.

As important as the rules, of course, were the tools: a smooth surface, even hoops, sturdy backboards, and a decent ball. As is often the case when new recreational sports receive considerable attention, companies began inventing and manufacturing products to meet the growing demand. One invention that has long pleased basketball players and fans is the improved seam-construction design in this patent awarded to George Pierce, who was acting as assignor to A. G. Spalding & Brothers. Prior to Pierce's unique stitching arrangement, the panels of hide on a basketball were tapered to polar points, which resulted in many inferior characteristics, such as the placement of a valve on the equatorial portion of the ball, and an altogether weaker structure. Pierce improved the design immeasurably by reshaping the panels.

How It Works

1. Panels are cut with abrupt, concavely curved ends.
2. Same panels include correspondingly curved projections at the centers.
3. Polar projections provide for a closure seam at one pole and an inflating valve at another.

NBA standards call for baskets that are 18 inches in diameter with the rim standing 10 feet above the floor. The ball itself must contain between 7.5 and 8.5 pounds of pressure.

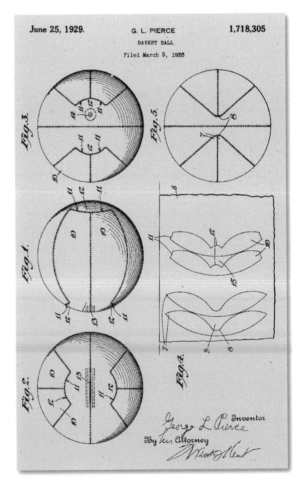

June 25, 1929.   G. L. PIERCE   1,718,305

BASKET BALL

Filed March 5, 1928

"What is claimed is:

1. A case for a game ball, composed of panels, each of which has length equal to a major part of the ball circumference, one continuously curved side edge, opposite, shorter, separately curved side edges, a convexly curved polar projection intermediate the separately curved side edges, and concavely curved ends to fit portions of the polar projections of another panel, the panels being connected together in pairs along said continuously curved side edges, and separately curved side edges of the panels of one pair being connected to similar edges of panels of another pair, and the concavely curved ends of the panels of one pair being connected to the polar projections of panels of another pair ..."

CORNFLAKES

Patent Name: Cereal Food and Process of Production
Patent Number: 1,321,754
Patent Date: November 11, 1919
Inventor: John L. Kellogg, of Battle Creek, Michigan

What It Does

Cornflakes are one of the world's most beloved breakfast cereals. Ingredients include milled corn, flavoring, iron, vitamin C, vitamin B6, vitamin B2, folic acid, vitamin A palmitate, vitamin B12, and vitamin D, all mixed, toasted, and flaked to make a reassuringly bland but filling morning meal (much enhanced nutritionally with the addition of milk).

Background

The story of cornflakes begins with two brothers and the mandates of the Seventh Day Adventist Church at the turn of the twentieth century. Dr. John Harvey Kellogg administered a health reform sanitarium founded by his church in Battle Creek, Michigan. John was a stern and educated doctor, obstinately devoted to the tenets of his religion, who ruled his patients, staff, and brother with an iron fist. Less educated and less interested in fire and brimstone, his

"Best to you each morning."

brother Will was a general maintenance worker and handyman around the place.

Will also became the cook, and his breakfast of choice consisted of hot cereals of grain, oat, or corn. In 1895, the morning after Will accidentally left some cereal on a hot stove, brother John discovered the flaky substance that resulted. He tasted it, deemed it palatable, refined it, and the first flaked breakfast cereal was born. Cold cereal was a new thing back then, but the Doctor blessed it and served it to the guests at the sanitarium. He believed the cereal would minimize his patients' desires to masturbate, a punishable no-no at the sanitarium.

As grain foods started turning profits, Dr. Kellogg attracted competition. His neighbor, C.W. Post, joined in the frenzy of moral cleansing by marketing Postum and Grape Nuts out of his own sanitarium. Meanwhile, Dr. Kellogg's own grain food company was losing money and the sanitarium was registering financial losses as well. In 1905, Will asserted his own business sense and convinced his brother to establish a new company to make cornflakes. The company was incorporated in 1906 and placed under Will's management. When the doctor was off on a trip, Will aggressively purchased stock from his fellow stockholder, making himself company president and putting his signature on the cereal box that now memorializes him.

Today, the company projects more than $9 billion in annual sales. It makes its products in nineteen countries, sells them in more than 150, and still exhibits some of its early vegetarian roots. Brand names the company markets include Kellogg's, Keebler, Pop-Tarts, Eggo, Cheez-It, Nutri-Grain, Rice Krispies, Special K, Kashi, and Morningstar Farms, which makes a variety of meat alternatives.

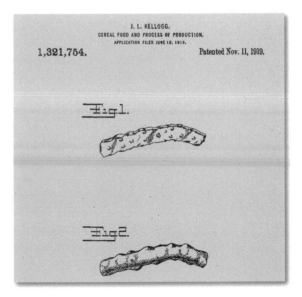

J. L. KELLOGG.
CEREAL FOOD AND PROCESS OF PRODUCTION.
APPLICATION FILED JUNE 10, 1919.

1,321,754.   Patented Nov. 11, 1919.

Fig.1.

Fig.2.

How It Works

1. A quantity of grain is cooked long enough so that its starches become dextrinized.

2. One method of cooking commonly used is a steaming conveyor on which the grain is propelled through pressurized steam.

3. After it has sufficiently cooked and partially dried, flavoring ingredients may be added.

4. The cereal is then toasted to give the flake its crispiness.

In the Inventor's Words

"My invention relates to the production of a ready-to-eat cereal food from any suitable grain, such as corn, rice, oats, wheat, or barley.

"My invention consists briefly of a cereal food made and the process of making the same in the form of separate shreds, flakes, or particles of other form…"

DRIVE-IN MOVIE THEATER

Patent Name: Drive-In Theater
Patent Number: 1,909,537
Patent Date: May 16, 1933
Inventor: Richard Hollingshead, Jr., of Riverton, New Jersey

What It Does

Provides a novel means of outdoor entertainment by projecting a movie on a large, open-air screen accompanied by personal listening devices so that people in parked cars can both see and hear films.

Background

As a young man selling auto parts at his father's shop, Richard Hollingshead began experimenting with a

little idea he had. He aimed a 1928 Kodak projector at a sheet stretched between two trees, and positioned a radio behind the sheet for sound. He even raised the front ends of cars on blocks for better viewing. Eventually, Hollingshead honed his ideas into a vision so precise as to include a method of keeping insects away from the film projector.

He broadly described his invention as "a novel construction to outdoor theaters whereby the transportation facilities to and from the theater are made to constitute an element of the seating facilities of the theater."

In other words, the audience is seated within the very means that brought them!

This aspect of Americana—the drive-in movie theater—combines our love affair with the automobile and our insatiable demand for entertainment. It has also helped jump-start many off-screen adolescent romances.

On June 6, 1933, less than a month after he was issued his patent, Hollingshead opened the first drive-in movie theater, in Camden, New Jersey, with a showing of *Wife, Beware.* The space could accommodate four hundred cars. Admission was twenty-five cents per person plus an additional twenty-five cents per auto—at this rate, a date could cost as little as seventy-five cents. Subsequent innovations, including in-car sound systems, increased the popularity of the drive-in theater; by the year 1958, just twenty-five years after the inventor opened the first one, more than four thousand drive-in theatres were scattered across the United States and Canada.

By the 1980s, however, the lure of the drive-in theater was overshadowed by the convenience of home entertainment systems and the development of large indoor multiplex theaters. Urbanization made the land that the drive-in theaters occupied more valuable than the service they provided; owners were pressured to sell to developers. By 1990, only around nine hundred drive-in theaters remained. While there is some interest in reviving them, drive-in movie theaters have become primarily the stuff of nostalgia, cultural symbols of a bygone era.

The drive-in theater with the largest capacity was the All-Weather Drive-In of Copague, New York, a 28-acre lot that could accommodate 2,500 parked cars. It included a monumental indoor viewing area that could seat 1,200.

How It Works

1. After dark, customers drive into prearranged slots in a large outdoor lot, slightly inclined and facing a screen.
2. A movie projector projects images onto a screen.
3. A central amplifier powers individual speakers temporarily attached to each car or, alternatively, centralized speakers project over a broad range.

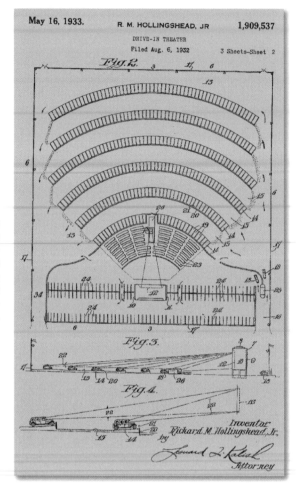

May 16, 1933.   R. M. HOLLINGSHEAD, JR   1,909,537

DRIVE-IN THEATER
Filed Aug. 6, 1932   3 Sheets—Sheet 2

Fig.2.

Fig.3.

Fig.4.

Inventor
Richard M. Hollingshead, Jr.
by
Leonard L. Kalish
Attorney

The second drive-in movie theater, the Shankweiler's Drive-In Theater in Orefield, Pennsylvania, was opened the year after Hollingshead's in 1934. It is now the oldest continuously operating drive-in.

4. Customers watch and listen from inside their cars.

In the Inventor's Words

The inventor included 20 specific claims in his patent application, 19 of which were amended variations on the first: "An outdoor theater comprising a stage, alternate rows of automobile drive-ways, and vertically inclined automobile stall-ways arranged in front of the stage, said stall-ways being adapted to receive automobiles disposed adjacent to each other and facing the stage;—said automobile stall-ways being at a vertical angle with respect to the stage such as will produce a clear angle of vision from the seat of the automobile, through the windshield thereof to the stage, free of obstruction from the automobiles ahead of it."

ETCH A SKETCH

Patent Name: Tracing Device
Patent Number: 3,055,113
Patent Date: July 23, 1959
Inventor: Arthur Grandjean, of Paris, France

What It Does   This educational toy provides hours of amusement as children manipulate the dials to create erasable line drawings.

Background   Imagine you are looking through a closed window that is covered in dust. You'd like to put your finger to the dust and scribble a pattern into it, but the dust is on the opposite side of the glass. By design, the inventive principles employed by the Etch A Sketch would allow you to draw such a pattern from your side of the glass. Arthur Grandjean invented his L'Ecran Magique (Magic Screen) in his garage in Paris, and showcased it at the 1959 International Toy Fair in Nuremburg, Germany. Two Americans representing the Ohio Art Company bought the rights to develop the invention, and the small toy company turned it into the now-classic silver-screened toy framed in red plastic with two white dials.

How It Works   An adherent material, such as aluminum powder, sticks to the underside of the screen's surface. Under the screen is a stylus that can be moved by two knobs at the bottom of the frame to move the powder aside and create patterns on the screen from below. One knob controls vertical movement of the stylus; the other, horizontal. Turning the knobs simultaneously can produce diagonal lines, and requires a degree of skill and coordination to accomplish a desired affect. To erase, simply shake the entire screen and the aluminum powder flows back into place.

1. A moveable stylus mechanism, (10) in patent illustrations, directly contacts the surface of the glass.

2. The stylus can be moved along two rods (11, 12), which are arranged at right angles to each other and are adapted to move in orthogonal directions by means of two cable systems (13, 14)

3. Turning the knobs (17, 18) engages movement of the cables, which in turn engages movement of the rods to which the stylus is affixed.

4. The stylus scratches a line of aluminum dust from the screen, creating a visible line.

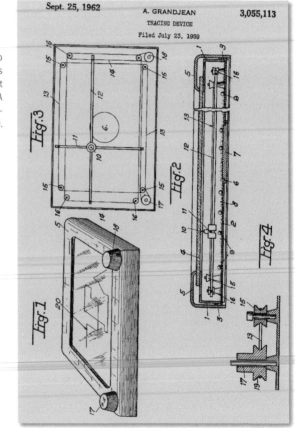

Sept. 25, 1962   A. GRANDJEAN   3,055,113

TRACING DEVICE

Filed July 23, 1959

Formed in 1908, the Ohio Art Company still makes and markets its most popular product, Etch A Sketch, which was introduced in 1960.

In the Inventor's Words

"This arrangement, which is capable of many applications for advertising purposes, for moving pictures, toys, testing means for instance, is constituted chiefly by means of a fluid-tight case provided in its upper part with a translucent surface of glass or the like material which is made opaque when the case is turned upside down, said case being partly filed with a pulverulent metallic or other material adapted to adhere then to said translucent surface and being provided inwardly with a removable tracing stylus which engages frictionally the inner surface of the translucent surface, so as to remove through its passage the pulverulent material adhering thereto and to form lines which are visible from the outside of the case and may be immediately wiped out through a mere turning of the case upside down associated with a shaking of the latter."

FENDER STRATOCASTER

Patent Name: Guitar
Patent Number: D169,062
Patent Date: March 24, 1953
Inventor: Clarence L. Fender, of Fullerton, California

What It Does   This superior design for an electric guitar features an ergonomic and attractive body, and is widely considered to be an electric guitar of legendary quality.

Background   Unlike acoustic guitars, which depend on hollow-body resonance, electric guitars have solid bodies and depend on electromagnetic pickups and amplifiers

to produce their unique and powerful sound. Though not a player himself, Clarence "Leo" Fender was an electric guitar aficionado who possessed a keen understanding of how the instrument worked. With Old World attention to craft, Fender made his guitars into masterpieces. For his New World business sense, he was sometimes called the Henry Ford of the electric guitar—he regularly updated his designs for the needs of an individual musician. His models were based on a variety of different guitar designs with funky names like the Broadcaster, the Telecaster and, most famously, the Stratocaster.

Repeatedly acquiring new Stratocasters, Jimi Hendrix must have been one of Leo Fender's most devoted customers. The left-handed musician was known to burn and smash his guitars during his electrifying performances. Other Strat-playing stars include:

· Dick Dale
· Buddy Holly
· Stevie Ray Vaughn
· Eric Clapton
· Chrissie Hynde of the Pretenders

Featuring an attractive and ergonomic design, the Stratocaster came to include components that rival electric guitars lacked. A bolt-on neck allows for speedier manufacture and if the neck happens to warp, the entire instrument need not be trashed. Three electromagnetic pickups are strategically placed to transfer the vibrations at bass, mid-line, and lead positions into electric signals. Covers for the pickups help eliminate undesired feedback, and higher notes become suddenly more accessible. The Fender Stratocaster is considered to make a unique new sound that helped launch the psychedelic and surf rock genres. Indeed, the Stratocaster was the instrument of choice for electric surf guitar guru Dick Dale and the left-handed rock virtuoso Jimi Hendrix.

How It Works

1. The body is usually constructed of a hard wood, such as maple.
2. Bright car paint may be applied to pleasing effect.
3. Electric components are added to the construction.
4. Strings are attached and tuned.
5. The body plugs into an amplifier.
6. The musician plays an altogether superior guitar.

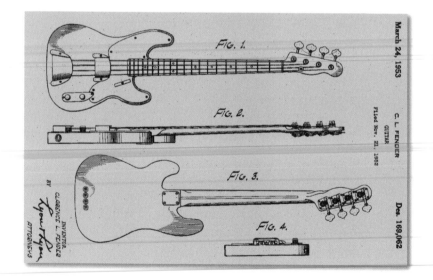

| In the Inventor's Words | "Be it known that I, Clarence L. Fender, a citizen of the United States, residing at Fullerton, in the county of Orange and State of California, have invented a new, original, and ornamental Design for a Guitar, of which the following is a specification, reference being had to the accompanying drawing forming a part thereof. |

No design changes were made to the Stratocaster until the 1980s, and even then, the changes were quite subtle.

Figure 1 is a top plan view of a guitar embodying my new design;

Figure 2 is a side elevation view thereof;

Figure 3 is a bottom view thereof; and

Figure 4 is an end elevation view thereof.

I claim:

The ornamental design for a guitar, as shown."

JUKEBOX

Patent Name: Coin-Actuated Attachment for Phonographs
Patent Number: 428,750
Patent Date: May 27, 1890
Inventors: Louis Glass and William S. Arnold, of San Francisco

What It Does The jukebox is a coin-operated music machine, usually found in restaurants or other public establishments, which allows a customer to select from an inventory of single songs and play them through public speakers.

Background Louis Glass and William S. Arnold pioneered a cultural phenomenon when they installed and demonstrated a coin-operated phonograph in San Francisco's Palais Royale Saloon on November 23, 1889. It was a far, and quiet, cry from what we call a jukebox today: The nickel-in-the-slot machine contained only one selection that a maximum of four people were able to listen to simultaneously through "hearing tubes." Amplification had yet to be invented.

The days of hip-shaking juke joints were still far off. The machine appeared decades before radio had hit the scene and just twelve years after Edison's phonograph was invented. It would be

Early jukeboxes spun platters, though today's models feature CDs.

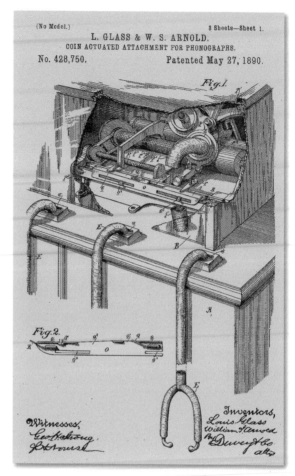

(No Model.) 3 Sheets—Sheet 1.

L. GLASS & W. S. ARNOLD.
COIN ACTUATED ATTACHMENT FOR PHONOGRAPHS.

No. 428,750. Patented May 27, 1890.

Fig.1.

Fig.2.

Witnesses,
Geo. Strong.
J. B. Morris.

Inventors,
Louis Glass
William S. Arnold
By Dewey & Co.
atty

Time magazine first printed the term Juke Box in 1939—an adaptation of the southern colloquialism "jook" which means "to dance."

a couple of decades before the phonograph became a regular feature in the household. Its timing made the nickel-in-the-slot novelty a huge hit. Patrons enjoyed music, proprietors attracted extra revenue, and a nascent music-recording industry inadvertently found one of the earliest ways to market popular songs.

As amplification did away with the earphones, and song capacity increased, chic designs came to give the jukebox a sense of style. Part record player, part vending machine, part furniture, the jukebox established itself as a valuable and attractive component of nightlife culture. While the jukebox has gone through countless mutations, the concept of dropping a coin in a slot to hear a good tune in public has remained perennially popular.

1. A mechanism for rendering the phonograph alternately operative and inoperative consists of a sliding piece that compresses the listening tubes.
2. A rock-shaft consists of two arms, one engaging pressure on the sliding piece, one projecting into the coin chute.
3. The coin is dropped into the chute, striking one arm of the rock-shaft.
4. The other arm releases pressure from the sliding piece, allowing a recording to be heard through the listening tubes.

In the Inventor's Words

"Our invention relates, generally, to the class of devices designed to be operated by a suitable coin deposited properly, and especially to an attachment of this class intended to be operated in connection with a phonograph.

"Our invention consists in the novel constructions and combinations hereinafter fully described, and specifically pointed out in the claims.

"The object of our invention is to provide a suitable device by which the phonograph may be exhibited and heard by any one upon the deposit of a suitable coin."

PEACH

Patent Name: Peach
Patent Number: PP15
Patent Date: April 5, 1932
Inventor: Luther Burbank

What It Does

A member of the Rosaceae family (of which the rose belongs), the edible delicacy known as the peach is as sweet as its flowered cousin is pretty. The fruit was first cultivated in China.

Background

Who doesn't love a peach? Or a plum? Or a potato, for that matter? For these and other healthy foods, we owe a great debt to Luther Burbank, born in 1849. Obviously, Burbank did not "invent" the fruits of Mother Nature, but he did pioneer the field of horticulture—

the science of cultivating plant life—to bring many edible varieties and create viable year-round markets for farmers everywhere. He experimented by grafting the seedlings of various plants onto other fully developed plants, often obtaining superior hybrids. More than eight hundred plant strains and varieties were created by this singular man.

Helping Ireland to recover from the black rot epidemic that wiped out its potato crops and caused massive starvation, Burbank developed a new strain of potato, selling the seeds for this forerunner to the Idaho potato for a meager fee. His earnings allowed him to move from Massachusetts to California, where he continued to work in horticulture—but he received no patents in his lifetime, as plant strains were not yet patentable.

After creating many new strains of fruits, vegetables, and other plants, Burbank died in 1926, a mere four years before the Plant Patent Act was instituted. Burbank received several patents, including this one, posthumously. The Act, meant to provide financial incentives to farmers and horticulturists, has itself borne much fruit, as Thomas Edison predicted in supportive testimony for the new legislation: "This (bill) will, I feel sure, give us many Burbanks.'

How It Works

Burbank's strain of peach is grown from a tree that is larger than other peach tree varieties of the same age. Its branches are stout and its twigs contain medium-to-large dormant fruit buds before bearing the fruit.

In the Inventor's Words

Luther Burbank

"This new variety of peach has resulted from years of experimenting with a definite objective in mind, that is, to produce a satisfactory yellow freestone peach which ripens halfway between the ripening periods of the known varieties, the June Elberta and the Early Elberta. It is similar to the Hale peach except that it has a large pit. It blood and seed are similar to the Muir, but the fruit is more golden in color. It is a stronger growing tree than the Valient and is not subject to peach curl and disease (Bacteria impruni) as is the last named variety. This new variety produces a very large fruit which averages about one-half pound. Its golden color with maroon shadings modified by a grayish pubescence, adds to its effectiveness in size. Although the skin of the fruit is thin and tender, tests have proven it to be a remarkable shipper; coupled with its great size, impressive coloring, excellent quality, and being a freestone, it represents an outstanding commercial peach. When cut in half, a pleasing modified apricot yellow flesh is disclosed which has a peach red tinge near the pit."

PHONOGRAPH

Patent Name: Improvement in Phonograph or Speaking Machine
Patent Number: 200,521
Patent Date: February 19, 1878
Inventor: Thomas A. Edison, of Menlo Park, New Jersey

What It Does   The phonograph was the first machine to record and play back sound, using principles similar to those of the telephone, along with an amplifying mechanism. It was eventually modified only to play sound recorded elsewhere professionally onto vinyl disks.

Background   A machine that can reproduce sound when a stylus is applied to the grooves in a disk, the phonograph stands among Edison's greatest inventions. Its intro-

duction sent Edison riding ever higher on the waves of international fame while it single-handedly spawned the birth of the recording industry. Not only does his patent describe the predecessor to the vinyl LP and record player, it describes similar features of a tape recorder. The invention could record and transcribe sound. It began as an offshoot of some experiments Edison was conducting in 1877 to improve his telephone and the diaphragm it used to transmit the sound of the voice. One inventive instinct led to another, and one can only wonder at the joy and satisfaction Edison must have felt when his stylus and tinfoil cylinder reproduced his own recorded voice intoning "Mary had a little lamb."

How It Works   To Record:
1. A cylinder (a) has a helical indenting groove cut from end to end. (Edison offered an estimate of ten grooves per inch.)
2. The material (tinfoil in Edison's invention) to be indented is placed over the cylinder and secured to a shaft (X).
3. At one end, the shaft has a thread cut into it at about ten threads per inch.
4. A pillar bearing (P) also has a single thread cut into it.

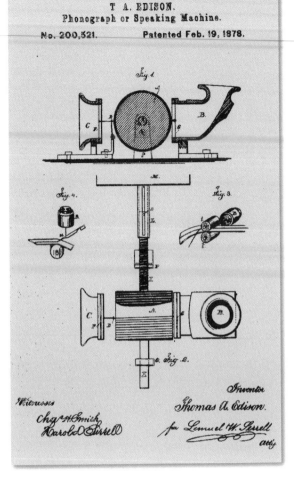

T. A. EDISON.
Phonograph or Speaking Machine.
No. 200,521. Patented Feb. 19, 1878.

5. A tube (L) can be fitted over the shaft to be rotated by means of clockwork or another power source (M).

6. A speaking tube (B) contains a diaphragm with an indenting point secured at its center.

7. The material passes between the pillar bearing and the indenting point as the machine is powered, registering the vibrations being indented upon it.

To Play:

1. When the sound has been imprinted, it can be reproduced through the listening tube (C).

2. Like the speaking tube, the listening tube contains a diaphragm, but instead of an indenting point, a light spring with a stout point is attached at its center.

3. When the cylinder rotates, the spring is set in motion to correspond with the indentions and thereby produce a facsimile of sound.

To Mass Produce:

1. The recording may be stereotyped by plaster-of-paris molds.

2. New sheets of foil can be pressed from these "master" molds.

In the Inventor's Words

"The invention consists in arranging a plate, diaphragm, or other flexible body capable of being vibrated by the human voice or other sounds, in conjunction with a material capable of registering the movements of such vibrating body by embossing or indenting or altering such material, in such a manner that such register-marks will be sufficient to cause a second vibrating plate or body to be set in motion by them, and thus reproduce the motions of the first vibrating body."

RADIO

Patent Name: System of Signaling
Patent Number: 725,605
Patent Date: April 14, 1903
Inventor: Nikola Tesla, of New York, New York

What It Does

The radio broadcasts distinct electrical impulses through radio waves that carry sound over distances and can be heard when tuned at certain frequencies.

Background

Nikola Tesla rivaled his former employer Thomas Edison in a variety of electricity-based innovations; using a motor he invented and the energy of the Niagara Falls, he successfully demonstrated hydro-electricity; and he invented a high-frequency "Tesla coil," still used today in radio and television transmission equipment. Yet, while Tesla was a brilliant inventor, he was never wealthy.

All too often, he was robbed of the credit he deserved for his inventions. The radio is one such example, demonstrating that Edison would not be Tesla's only rival. In this case, Italian inventor Guglielmo Marconi demonstrated a radio voice broadcast in San Francisco in 1895; newspaper reports hailed the phenomenon a new invention. Never mind the fact that Tesla had demonstrated radio communication in Philadelphia two years earlier. Today, Marconi is still often recognized as the inventor of the radio. But Tesla's work speaks for itself: the radio, and the principles behind so many other technologies we employ today, would not be possible without Tesla's driving intellect.

With or without credit, Tesla continued experimenting and improving earlier inventions. The underlying principles of an earlier patent Tesla received for a system of transmitting electrical energy are the focus of this patent.

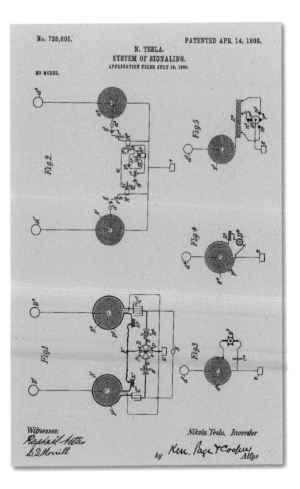

1. In figure 1, two spirally wound conducting coils (S^1, S^2) connected with their inner ends to elevated terminals (D^1, D^2).
2. Their outer ends connect to an earth-plate (E).
3. The two systems may have electrical oscillations impressed on them by energizing them through primaries (P^1, P^2) and regulated by inductances (L^1, L^2).
4. The two independent primary circuits send vibrations to the earth-plate and spread to a distance reaching a similar circuit-forming receiving station tuned in conjunction with the sending-station.

In the Inventor's Words

"Broadly stated, this invention consists in the combination of means for generating and transmitting two or more kinds or classes of disturbances or impulses of distinctive character with respect to their effect upon a receiving-circuit and a distant receiver which comprises two or more circuits of different electrical character or severally tuned, so as to be responsive to the different kinds or classes of impulses and which is dependent for operation upon the conjoint or resultant action of the two or more circuits or the several instrumentalities controlled or operated thereby."

INVENTOR PROFILE

Nikola Tesla (1856–1943)

Nikola Tesla

Nikola Tesla was born in Croatia to a Serbian Orthodox-priest father and an uneducated but highly intelligent mother. Tesla's greatest contribution to the field of electrical engineering was the discovery of the rotating magnetic field, which forms the basis of almost all alternating-current machinery.

Tesla was educated in Austria and at the University of Prague. While studying the Gramme dynamo, a direct-current generator that becomes an electric motor when reversed, he got the idea to begin experimenting with alternating currents. In 1882, Tesla went to work for the Continental Edison Company in Paris. While on assignment in Strasbourg, France, in 1883, Tesla built his first inductor motor, one of the most common types of alternating-current motors. It was the first polyphase AC motor—meaning, the coils in the motor were arranged in such a way that out-of-step currents energized the magnetic field, causing it to rotate at a predetermined speed. Such a discovery helped pave the way for a practical alternative to direct current. Alternating current was important because it could be more easily modified to suit a variety of situations.

Despite his highly technical way of thinking, Tesla was also a dreamer, prone to writing poetry when not working. True to his nature, in 1884, Tesla set out for America with a few poems and just four cents to his name. Upon arriving in America, Tesla worked for a brief time under the direction of Thomas Edison, but Edison was a strong proponent of direct current technology and had little interest in the young engineer's work. Tesla quit, frustrated, after only one year.

In 1887, Tesla started the Tesla Electric Company in New York City. There, he experimented with a variety of technologies, including the shadowgraph, which presaged the invention of X-rays in 1895 by Wilhelm Roentgen. In 1888, Tesla sold his suite of AC inventions—the alternating-current motor and the transformers and devices that went along with them—to George Westinghouse. In 1893, Westinghouse used Tesla's system to light the World's Columbian Exposition at Chicago. Tesla's motors were used in 1893 to power the Niagara Falls Power Project. Tesla's humble AC motor harnessed the great power of the falls to deliver electricity 22 miles away in Buffalo, heralding a new era in modern electricity.

Tesla's other inventions include the 1891 Tesla coil, a high-frequency coil that is still used today in radio and television transmission equipment. He also experimented with remote control and radar, and developed fluorescent lighting. In fact, Tesla earned over one hundred patents in his lifetime, but he died nearly destitute, alone in his room at the famed New Yorker hotel. However, his legacy lives on, and in 1975 he was inducted into the National Inventors Hall of Fame.

ROLLER COASTER

Patent Name: Roller Coasting Structure
Patent Number: 310,966
Patent Date: January 20, 1885
Inventor: La Marcus A. Thompson, of South Chicago, Illinois

What It Does This thrill ride, in its many variations, carries passengers along a track in small, individual, trainlike cars, over steep drops and around precipitous curves, providing terror and delight in equal measure.

Background In June 1884, the first contemporary commercial roller coaster ride, the Gravity Pleasure Switchback Railway, opened to the public at Coney Island, New York. The ride cost a nickel a person and, at its fastest, moved at a whopping six miles per hour. It consisted of a platform for patrons which included a ticket office and flat steel tracks nailed into wooden planks, the whole of which was suspended at varying heights by trestles. Like many innovations, the "roller coasting structure" represented a combination of evolving concepts of the time. Elsewhere, others were devising similar pleasure rides, including boat-slide chutes and inclined railways, as the novelty of large-scale amusement parks was quickly becoming an industry.

It has been reported that L. A. Thompson was a preacher whose initial intention in building the ride was to divert attention from Coney Island's beer gardens. Preacher he may have been, but entrepreneur he quickly became. After enjoying the immediate success of his invention, Thompson went on to build dozens more roller coasters of varying designs. By the early 1900s, he had employed as a chief engineer a man named John Miller, who began designing his own rides. Miller held over one hundred patents on roller coaster innovations and is today considered the father of the modern roller coaster. During the 1920s, roller coasters were all the rage, but following the Great Depression, the industry suffered enormously. In the four

Considerable advances in roller coaster technology have focused on improving safety while maximizing thrill. Opened in 2003 with a construction cost of $25 million, Cedar Point's Top Thrill Dragster in Sandusky, Ohio, has the largest drop—400 feet—and the top speed—120 mph—to date.

decades that followed, many more roller coaster rides were dismantled than built. It is only recently, with the rise of such super-parks as the Great Adventure chain, that the roller race has heated up again, as engineers and designers strive to create the biggest, steepest, fastest, and most tortuous ride yet.

How It Works

1. A parallel double track structure (B, B') is supported by a trestle system (C), and includes both ends situated at an equal height above the rest of the track.
2. From a starting point (D) of the outgoing track, a car gains enough momentum as it descends to reach the other side.
3. Here, the car is transferred to the return track (B') via a switch-track mechanism (E), and similarly moves back to its starting point.

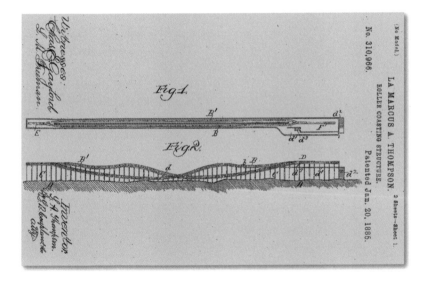

In the Inventor's Words

In August 2001, two Japanese men rode in a roller coaster marathon that lasted over thirty-five hours.

"This construction and arrangement affords a very enjoyable means for amusement and pleasure, the sensation being similar to that of coasting on the snow, with the difference that the conveyance runs on wheels and returns the passenger to the starting point without the necessity of having to walk up hill for a second ride."

ROLLER SKATES

Patent Name: Roller Skate
Patent Number: 906,281
Patent Date: December 8, 1908
Inventor: James Leonard Plimpton, of Boston, Massachusetts

What It Does

First fashioned in 1863, the four-wheel roller skate gave skaters more maneuverability than its in-line predecessors, which were modeled on ice skates. Though it had begun as a recreation for the aristocracy, roller-skating became a popular and widespread amusement.

Background

1. In figure 1, a shoe is strapped securely to the skate by straps.
2. The trucks of the skate are designed to allow for leveraging to one side or another under pressure, allowing the direction of the skater to curve slightly in a preferred direction.
3. Braking is performed by applying pressure down on the heel where a screw bolt (48) engages the brake mechanism (47).

How It Works

Originally, roller skates were a lot more like today's in-line skates, roughly replicating the blade of an ice skate. According to Mary Bellis, who hosts the inventor Web site inventors.about.com, a German ballet, *Der Maler oder die Wintervergnügen* (*The Artist, or Winter Pleasures*), used roller skates to mimic ice-skating in 1818. The following year in France, a patent was issued for a skate that had in-line wheels.

James Plimpton changed the "in-line" design with his 1863 invention of a rocking skate, mounting two pairs of wheels side by side to allow skaters more maneuverability. Skaters could lean one way or another to steer as opposed to lifting their feet numerous times in increments. The innovation was favorably received, and skating continued to grow in popularity. An enterprising businessman, Plimpton would soon establish

This patent model represents one of dozens of new roller skate innovations that followed Plimpton's original four-wheeler.

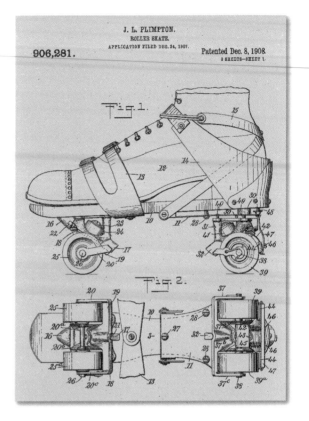

J. L. PLIMPTON.
ROLLER SKATE.
APPLICATION FILED DEC. 24, 1907.

906,281.

Patented Dec. 8, 1908.
2 SHEETS—SHEET 1.

a skating club in New York City. This invention represents an improvement on the first patent awarded to the person who is now known as the father of modern roller skating.

In the Inventor's Words

"This invention relates to and is dependent upon the construction, arrangement, and mode of operation of that class of guidable curved-running roller skates, as originated and first patented by me in the United States, January 6, 1863, and more particularly as improved and patented by me in England, August 25, 1865, and in the united States, June 26, 1866. In this class of guidable curved-running roller skates the rollers are applied to the stock of foot-stand of the skate so that the said rollers may be cramped or turned, so as to cause the skate to run in curved lines either to the right or left by the turning, canting, or tilting laterally of the stock or foot-stand."

SLINKY

Patent Name: Toy and Process of Use
Patent Number: 2,415,012
Patent Date: January 28, 1947
Inventor: Richard T. James

What walks down stairs, alone or in pairs, and makes a slinkity sound?
A spring, a spring, a marvelous thing. Everyone knows it's Slinky . . .
                                                    —TV Commercial Jingle

What It Does   The Slinky doesn't really do much—but it does it with
endless style. The twangy little spring toy can "walk"
downstairs, but mainly it just begs you to play with it.
"Kazhinnnnng."

Background   As a naval engineer in Philadelphia, Pennsylvania,
Richard James experimented with tension springs in
attempting to devise an antivibration system for sensitive equipment in ships. While he worked one after-
noon in 1943, a coiled set of springs fell to the floor.
The serendipitous accident revealed its unique potential outside of a ship—as a toy. To the engineer, it was
the intriguing expression of potential energy transfer-
able into kinetic energy. To the rest of the world, it

was a coiled spring that, once set into motion, became
instantly fascinating. It oscillated; it turned on its ends;
it walked itself down stairs.

He and his wife Betty worked on fine-tuning the
toy and during the 1945 Christmas season, they demonstrated the Slinky for the first time at a department
store. In ninety minutes, the entire stock sold out, four
hundred Slinkys for a buck apiece. Encouraged, the
couple established a company to mass produce the
toys. James developed a machine that could turn 80
feet of steel wire—the length of a standard Slinky—
into the desired coil in about ten seconds. James Industries was founded in 1956.

By the early 1960s, Slinky sales had slowed down.
Richard James went off to join a religious group in
Bolivia, leaving his wife with their six children and a
lot of debt. Like a coiled spring, Betty James bounced

back into motion. She took over at James Industries and saved Slinky from the shadows of obscurity, and the Slinky began to slink its way back into homes everywhere. Today more than 250 billion Slinkys have been sold worldwide.

How It Works

When set in motion on a stepped platform such as a stairway, a helical spring transfers energy along its length in a longitudinal wave. Coil by coil, and step by step, the whole spring descends end over end, as if it is somersaulting down one step at a time.

In 2001, the Slinky was inducted into the Toy Industry Hall of Fame. In the same year, it was named the Official State Toy of Pennsylvania.

The Slinky has been used as more than just a toy: Science teachers have used it as an educational aid, and in fighting the Vietnam War, soldiers tossed Slinkys into tree branches as makeshift radio antennae.

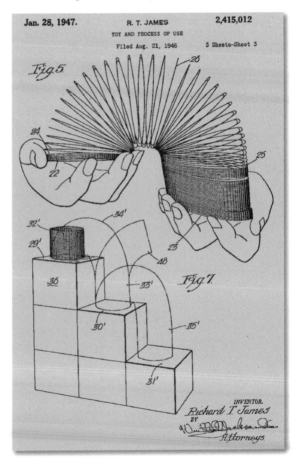

In the Inventor's Words

"A purpose of my invention is to provide a helical spring toy which will transfer its turns from one end to the other in an entertaining manner when it is bent into general semi-circular form and the ends are moved up and down.

"A further purpose is to provide a helical spring toy which will walk on an amusement platform such as an inclined plane or set of steps from a starting point to successive lower landing points without application of external force beyond the starting force and the action of gravity."

SNOWBOARD

Patent Name: Snowboard
Patent Number: 4,165,091
Patent Date: August 21, 1979
Inventor: Daniel E. Chadwick, of Rutherford, New Jersey

What It Does The snowboard combines the principles of the surfboard, skateboard, and snow skis to offer a recreational glide over snow-laden surfaces.

Background The popularity of snowboarding may seem like a recent phenomenon but people have been enjoying the sport, in one form or another, for a long time. Sled-surfing was destined to evolve, and the flat-bottomed snowboard is its current manifestation. The skills and coordination required for snowboarding are often compared to those of skateboarding; and many street-skaters answer the wintertime call of shredding the slopes.

How It Works This patent represents an early version of the snowboard, which very closely resembles a modified skateboard and borrows many of the same underlying principles. The board consists of an elongated main body having laterally spaced skis or runners mounted to the underside of the body at the front and rear. The runners, like wheels on a skateboard, are coupled together in order to maintain their parallel relation and the stability of the board. Each runner may have a centrally disposed stabilizer fin on its bottom surface.

1. A board is constructed from rigid plastic or wood.
2. Axially spaced pairs of runners with upward curving portions are situated on the bottom, near both ends.
3. The runners feature small central fins, which help stabilize the board while in use.
4. The runners are symmetrically located on opposite sides of the board's longitudinal axis and coupled together in parallel relation by transverse bars.
5. A middle transverse bar is vertically offset above the front and rear bars, and includes a flexible bracket, the upper end of which is secured to the bottom of the board.

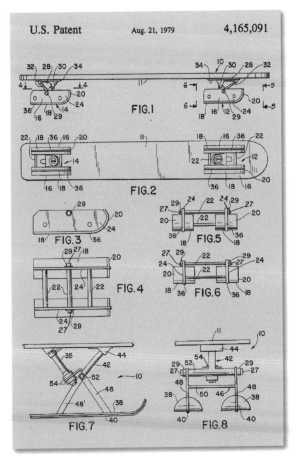

In 1999, Australian Darren Powell set a world record for snowboarding speed—125.459 mph.

U.S. Patent Aug. 21, 1979 4,165,091

FIG.1 FIG.2 FIG.3 FIG.4 FIG.5 FIG.6 FIG.7 FIG.8

In the Inventor's Words

"According to a preferred mode of operation the board is used by placing it at the top of a hill which preferably is packed hard with snow. The operator stands on the board and pushes off, whereupon both feet

are placed on the board, being located similar to their
location on a 'skateboard.' As the board slides over
the snow proceeding downhill, the operator may lean
to either side in order to manipulate the board in a
preferred direction. The resilient connection will aid in
manipulation of the board by the operator."

STATUE OF LIBERTY

Patent Name: Design for a Statue
Patent Number: 11,023
Patent Date: February 18, 1879
Inventor: Auguste Bartholde, of Paris, France

What It Does Offered as a gift by France to the United States, the
monumental Statue of Liberty, which sits in New York
Harbor, was designed to embody Liberty enlightening
the world.

Background She is a tourist attraction; she is a piece of art; she is a
beacon of welcome to America's immigrants enchanted
by the possibilities of a better life in a new land. One
of the single most photographed objects in the world,
Lady Liberty is a universal symbol of freedom. A gift
from France to commemorate the centennial of the
American Declaration of Independence, the statue was
conceived at a dinner party in Paris in 1865. Six years
later, in order to check out sites and get some ideas for
the monument, the French sculptor boarded a ship
bound for America. Before it docked, Bartholde had
selected Bedloe's Island as the site and had even drawn
up some sketches.

In 1876, the statue's torch-bearing arm was demonstrated in Philadelphia in honor of the Centennial of
the Declaration of Independence; the rest of the statue
had not yet been finished. The sculptor conceived
of the whole of his work as consisting of an inner
supporting framework of steel covered by plates of
thin copper. Alexandre Gustav Eiffel, who later built
another famously photogenic monument, oversaw
construction of the inner framework. Completed in

France in July, 1884, the statue arrived the following year by boat, like so many European emmigrants of the day. The vessel carried Lady Liberty in more than 300 pieces packed in some two-hundred-plus crates. The Statue was inaugurated by President Grover Cleveland on October 28, 1886. In 1903, these verses of Emma Lazarus were inscribed at the statue's base:

Give me your tired, your poor,
your huddled masses yearning to breathe free,
the wretched refuse of your teeming shore.
Send these, the homeless, tempest-tost to me.
I lift my lamp beside the golden door!

Joseph Pulitzer helped obtain contributions to complete the statue's base in April 1886. Benefit funds for Lady Liberty were also raised by theatrical events, art exhibitions, auctions, and prizes.

DESIGN.

A. BARTHOLDI.
Statue.

No. 11,023.   Patented Feb. 18, 1879.

LIBERTY ENLIGHTENING THE WORLD.

An inner framework of steel is sheathed in multiple molded plates of thin copper.

In the Inventor's Words "The statue is that of a female figure standing erect upon a pedestal or block, the body being thrown slightly over to the left, so as to gravitate upon the left leg, the whole figure being thus in equilibrium, and symmetrically arranged with respect to a perpendicular line or axis passing through the head and left foot. The right leg, with its lower limb thrown back, is bent, resting upon the bent toe, thus giving grace to the general attitude of the figure. The body is clothed in the classical drapery, being a stola, or mantle gathered in upon the left shoulder and thrown over the skirt or tunic or under-garment, which drops in voluminous folds upon the feet. The right arm is thrown up and stretched out, with a flamboyant torch grasped in the hand. The flame of the torch is thus held high up above the figure. The arm is nude; the drapery of the sleeve is dropping down upon the shoulder in voluminous folds. In the left arm, which is falling against the body, is held a tablet, upon which is inscribed "4th July, 1776.""

STATUE STATS

· Height from base to torch: 151'1" (46.05m)

· Length of right arm: 42' (12.8m)

· Index finger: 8' (2.44m)

TELEVISION

Patent Name: Cathode Ray Tube
Patent Number: 2,139,296
Patent Date: December 6, 1938
Inventor: Vladimir K. Zworykin, of Philadelphia, Pennsylvania, assignor to Radio Corporation of America

What It Does The television is the bearer of news, the provider of entertainment, and a medium for commercial advertising. Perhaps more than any other medium, it is responsible for the dissemination of popular culture, for better or for worse.

Background The story of the invention of television would itself make for compelling TV drama. To this day, there remains debate as to who should receive credit. But

Within seconds after it was introduced, the TV set moved from the living room into the kitchen. . . and the bedroom, and the kid's room, and finally even the car . . .

like so many complex innovations, the TV was an awesome creation just waiting to happen, a keystone balanced atop previous ingenious ideas. One of the innovations key to its foundation—the cathode ray tube—was invented in 1897 by a German scientist named Karl Braun. A specialized vacuum tube that can transfer refracted electron beams into images, the cathode ray tube is to this day used in TVs, video cameras, and many other image-related technologies.

About a decade after Braun created his cathode ray tube, a Russian physicist named Boris Rosing began tinkering with the idea, enlisting the help of his student Vladimir Zworykin. While Rosing was first to use Braun's tube in an attempt at producing television images, he could only manage to transmit crude shapes. Rosing's experiments were cut short when he disappeared sometime during the 1917 Bolshevik Revolution. Eventually, his student would enjoy another chance at realizing the dream.

Zworykin left his home country to study the relatively new X-ray technology in Paris, eventually traveling to the United States in 1919, where he got a job at a Westinghouse lab in Pittsburgh, Pennsylvania. Here, and later with significant funding from RCA, he continued to experiment with TV technology, making significant improvements to the cathode ray tube for the purpose of generating electronic images.

RCA would later fund another inventor who made a legitimate case that he had already invented and patented the basic components of electronic television. A farm boy from Utah, Philo Farnsworth was the first to transmit an image made by sixty horizontal lines on a TV screen. In 1927, he filed for his first television patent, number 1,773,980. RCA sent its Russian engineer to meet the Utah farm boy to obtain information on perfecting their television. Farnsworth's company hosted the Russian, thinking it might be able to obtain a valuable contract from the much larger RCA.

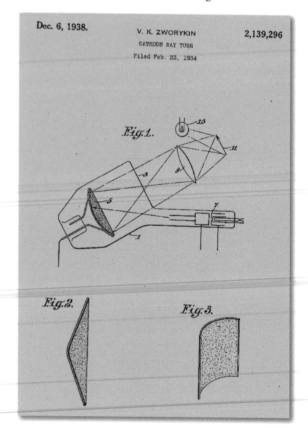

Dec. 6, 1938.

V. K. ZWORYKIN

CATHODE RAY TUBE

Filed Feb. 23, 1934

2,139,296

Fig.1.

Fig.2.

Fig.3.

After arduous legal wranglings and patent stand-offs, RCA conceded to pay one million dollars for the rights to Farnsworth's patents. Though this fact is probably not something RCA would advertise in its corporate history, it still does not answer with total certainty the question of who was the true television inventor, if there is only one, and debate on the subject lingers. Both Zworykin and Farnsworth had workable television systems by the year 1930, and both are credited as "the inventor."

What is certain: television has made the world smaller and has profoundly altered social landscapes everywhere.

How It Works

1. Within the cathode ray tube, electrons are directed at a photosensitive target, and are refracted toward its flat surface, which comprises the television screen.
2. When electrons strike the phosphorescent coating along the flat surface of the tube, light is emitted.
3. Meanwhile, around the outside of the tube, electromagnetic coils deflect the electron beam so that it scans across the screen, usually in a horizontal pattern.
4. When the scanning is maintained at a rapid pace, the light on the screen forms images that appear steady and continuous to the human eye.

In the Inventor's Words

"A cathode ray transmitting tube as disclosed in my copending application is constituted by an evacuated container having at least one transparent wall, a planar photo-sensitive cathode of the mosaic type, and an electron gun for directing a cathode ray toward and against the cathode.

"When such a device is utilized in a television transmitter or in an ultra-microscope of the type disclosed in the aforesaid application, it is necessary to interpose an optical system, constituted by one or more lenses, between the object and the photo-sensitive electrode for the purpose of forming upon the said electrode an optical image of the object."

RESOURCES

THE THREE TYPES OF PATENTS

Three types of patents are awarded by the United States Patent and Trade Office. The most commonly issued patent, by far, is the utility patent. The other two types are plant patents and design patents.

1. Utility Patents

Utility patents are issued to any invention that can be considered a useful process, machine, method, or composition of matter. Utility patents also cover the development of new uses for any previously existing inventions in any of these categories, as well as improvements upon inventions that are already around. The good thing, as far as inventors are concerned, is that patents don't have to be more useful, original, or appealing than anything else on the market; they simply have to have a baseline of usefulness about them. The key to being granted a utility patent lies in the functionality of the invention. As long as something can be considered "new and nonobvious" (in the words of the Patent Office), it will most likely be awarded a patent.

The extreme majority of patents represented in this book, as in life, are utility patents. In recent years, the field of biotechnology has broadened the definition of the utility patent. Patents are now granted to cover forms of life developed in a laboratory through genetic engineering. New breeds of mice, never-before-seen strains of corn, and many types of microorganisms are examples of the utility patent. Utility patents expire twenty years after being issued.

2. Plant Patents

The Plant Patent Act of 1930 gave intellectual property protection to inventors of new kinds of plants. Plant patents are granted to anyone who discovers or invents and asexually reproduces a distinct and new variety of plant. This includes mutants, hybrids, or newly found seedlings. An asexually reproduced plant has been propagated not from a seed but from cutting, grafting, inarching, or another similar process. Plant patents do not apply to tuber-propagated plants, such as the Irish potato or Jerusalem artichoke. When submitting a patent application, the inventor of the new plant must include a detailed drawing of the specimen; if the new plant is distinguished by a new color, the color must be shown in the drawing. Plant patents are often issued to individuals who have created new plants for a variety of reasons: To create trees that bear more fruit, or a new kind of fruit; to create plants (or crops) that are more resilient to disease; to cultivate plants with medicinal properties; or simply to bring into existence a new kind of aesthetically pleasing flower. Like utility patents, plant patents expire after twenty years.

3. Design Patents

Design patents are granted to anyone who has come up with a new and "nonobvious" ornamental design for a manufactured item. One such example in this book is the Fender Stratocaster. The design patent only covers the appearance of an item, not its functionality, and prevents anyone else from making something that looks the same as the product that has been issued the design patent. Approximately one design patent is issued for every 10 utility patents out there, making them relatively rare in the patent world. The interesting thing about design patents is that they only cover the cosmetic appearance of a thing—the patent has nothing whatsoever to do with the item's function. As a result, when someone files a design patent application, the entire "invention" can only be illustrated through a drawing or other rendering; it is not considered valid if it is described in words. Design patents are sought out when the appearance of a thing will substantially influence that item's commercial success. So, for example, computer icons are often patented; however, the parts inside of a computer, which aren't normally seen, cannot be issued design patents, because it isn't the shape—or design—of the parts that makes them functionally useful. Design patents expire after fourteen years.

PATENTS VERSUS TRADEMARKS

Patents and trademarks form the basis of the work of the U.S. Patent and Trademark Office. Both protect holders of certain types of intellectual property. However, the exact function of the each is distinct.

A patent is a document that protects an invention, usually a product, plant, or design (see "The Three Types of Patents"). When the inventor is issued the patent, he or she receives certain protections that allow others to use, analyze and/or modify the invention in the name of the advancement of science, while still granting the inventor any proceeds, profits, or other benefits that may arise from the invention.

On the other hand, there are trademarks: Bubble Wrap; Post-It Notes; Etch A Sketch; Velcro. Each of these creations is revolutionary in its own way; each has its story to tell; and each is trademarked, which is why you see the little circle with the "r" after each name. That symbol means "Registered Trademark." The success of these creations relies on brand-name recognition—and on successfully establishing the fact that the product is identifiable, distinct, and superior. While patents protect the duplication of a product by another manufacturer for a limited time, trademarks rely on consumer devotion to the original product, through marketing its name, next to which other similar products then become measured. Unlike patents, a trademark is valid for as long as the product is in use.

A trademark is simply a word, phrase, symbol, or design that appears on the product being sold and which distinguishes itself from other words, phrases, symbols, or designs in the marketplace. A service mark, yet another type of trademark, is used by companies that provide a service instead of a product. A service mark is also represented by a word, phrase, or symbol. Another word for trademark is "brand name," Companies work hard to make their trademarks distinctive. A strong trademark may become so embedded in the public's consciousness, in fact, that it virtually replaces the real name of the product. The above listed products serve as outstanding examples. One typically doesn't ask for "sealed-air cushioning material" when "BUBBLE WRAP" is so ingrained into the lexicon. "Kleenex" and "Xerox" offer other similar examples. These aren't vocabulary words at all, but trademarks or brand names. Same goes for Coke, the nickname for Coca-Cola. But while the Coca-Cola name is trademarked, the secret formula used to create its flavor is patented.

In other words, a trademark really protects a marketing idea—the image of a product that a company is trying to sell. When a company issues a trademark, it wants to make sure that the consumer doesn't confuse its product with some other company's product. So the trademark protects the company's image or reputation and in doing so, helps it build brand loyalty and stay in business. Trademarks have become increasingly valuable as the global marketplace has expanded. In a market flooded by goods from all over the world, companies are challenged to work harder to establish and protect their trademarks.

HOW TO OBTAIN A PATENT

The first thing you want to do before applying for a patent is to make sure that your idea hasn't already been patented. You can check on this by conducting a literature review at your local university library, or by searching on the U.S. Patent and Trademark Office Web site (http://www.uspto.gov/). While you're putting the final touches on your invention, be secretive! It's a good idea to keep records and documents of all the work, purchases, and other data that went in to the making and refining of your idea, just in case someone else comes up with a similar idea before you apply for your patent. That way, if similar invention is unveiled around the same time, you can prove that you indeed had the idea first.

If you're really anxious that news of your brilliant idea is going to leak out, you can even (for a fee) make use of a U.S. Patent and Trademark Office service called the Disclosure Document Program. Under this program, the Patent and Trademark Office will keep a record of your idea for two years. This doesn't in any way protect your idea; it's merely a more systematic way of establishing evidence that the concept came from you first, should anyone ever try to challenge you. In the United States, you have a yearlong grace period in which to apply for a patent after you make your idea public, should you choose to do so before you are granted a patent. But beware: no other country offers a grace period, and if you make your idea public, either by writing about it in a journal or by manufacturing and selling it, you forfeit your right to apply for a patent in other countries.

Once you're pretty sure that your invention is "new and nonobvious," you can begin the patent application process. There are three main components of any patent application: (1) A written document that specifies the details of the invention and includes an oath of declaration; (2) An accompanying drawing of the invention, if that's applicable; and (3) the filing fee. The U.S. Patent and Trademark Office calls the patent application a "complex legal document" that is best left to patent attorneys or agents. However, you can apply for one on your own if you're willing to fulfill the lengthy and exacting requirements. The USPTO even specifies that the paper you submit your exhibits on be "flexible, strong, smooth, nonshiny, durable and without holes."

Once you've attended to the details, the bulk of the application can be found in the Specification section. This section requires a full, concise, clear, and exact description of the invention that is explicit enough so that another person could construct or replicate it. Then comes all-important Claims section. Here, you must make as strong a case as possible for the relevance and usefulness of your invention, citing the aspects that make it distinct. The wording here is extremely important, and strongly influences whether your patent will be granted.

Next is the Drawing section, which should include various views of the invention, if applicable. Finally, you must be sure to pay the applicable fees. Current filing fees are as follows: utility patent filing fee, $770; design patent filing fee, $340; plant patent filing fee, $530. If your patent is granted, congratulations! But be prepared to pay even more fees. The issuance fee for a utility patent is $1,330, for a design patent, $480, and for a plant patent, $640. You will also have to pay various maintenance fees at $3 1/2$ years, $7 1/2$ years, and $11 1/2$ years after issuance of the patent. The maintenance fees grow incrementally as the years go by, but if your patent has proved useful to society, you'll surely be more than rich enough to afford them!

USEFUL RESOURCES

You can search for U.S. Patents in the Search Room of the Patent and Trademark Office in Crystal Plaza, 2021 Jefferson Davis Highway, Arlington, Virginia. Or, you can search for U.S. patents on the World Wide Web at: http://www.uspto.gov/patft/index.html. You can also call the USPTO's General Information Services Division at 800-PTO-9199 or 703-308-4357.

In lieu of traveling to Virginia to manually search for patents, you can make a trip to your nearest Patent and Trademark Depository Library (PTDL). Every state as well as the District of Columbia and Puerto Rico have PTDLs, where you can search patent and trademark collections. To find the nearest PTDL, visit http://www.uspto.gov/web/offices/ac/ido/ptdl/ptdlib_1.html

Online instructions for filing patent applications can be found on the United State Patent and Trademark Office, at these addresses:

A Guide to Filing a Utility Patent Application
http://www.uspto.gov/web/offices/pac/utility/utility.htm#intro

A Guide to Filing a Design Patent Application
http://www.uspto.gov/web/offices/pac/design/index.html

General Information About 35 U.S.C. 161 Plant Patents
http://www.uspto.gov/web/offices/pac/plant/index.html

You can also file for some applications online, such as utility patents and provisional applications, using the USPTO's electronic filing system. You cannot apply online for design or new plant applications.

Electronic Filing System
http://www.uspto.gov/ebc/efs/index.html

There are lots of fun sites out there dedicated to patent news, laws, history and information. Here are some good ones:

Invent Now Inventors Hall of Fame
A fun, nicely designed site whose aim lies in "fostering the inventive spirit in all of us."
http://www.invent.org/index.asp

Patent-FAQ
This list of Frequently Asked Questions assembled by Reg. U.S. patent agent David Kiewit is a treasure trove of information regarding patent applications and intellectual property in general.
http://www.patent-faq.com/

The Lemelson-MIT Program at the Massachusetts Institute of Technology
This site is dedicated to celebrating inventors and inventions. Their Web site includes a handbook for inventors. You can find it at:
http://web.mit.edu/invent/h-main.html

If you're serious about patent research, you can pay a fee to become a subscriber to the Delphion Research Intellectual Property Network, which allows you to search international and U.S. patents. Visit the site at:
http://www.delphion.com/

For comprehensive information about patent laws, visit the American Patent and Trademark Law Center at:
http://www.patentpending.com/.

BIBLIOGRAPHY

Amirani, Amir. "Sir Alec Jeffreys on DNA Profiling and Minisatellites," *Science Watch*, 1996.

Bacon, Tony. *The Ultimate Guitar Book*. New York: Alfred A. Knopf, 1997.

Bardey, Catherine. *Lingerie: A History and Celebrations of Silks, Satins, Laces, Linens and Other Bare Essentials*. New York: Black Dog and Leventhal, 2001.

Bowler, R. M., PhD, MPH and J. E. Cone, MD, MPH. *Occupational Medicine Secrets*. Philadelphia, PA: Hanley & Belfies, 1999.

Bullock A., and R. B. Woodings, eds. *20th Century Culture*. New York: Harper & Row, 1983.

"Chili Peppers and Endorphins," *The Veiled Chameleon*, April 21, 2003, http:// www .veiled-chameleon.com/archives/000042.html (accessed November 11, 2003).

Folkhard, Claire, ed. *2003 Guinness Book of World Records*. New York: Bantam Books, 2003.

Folkhard, Claire, ed. *2004 Guinness Book of World Records*. New York: Time, Inc., 2003.

"Genetic Fingerprinting," *The Science Show*, Radio National. Transcript of Broadcast, September 21, 2002.

Hancock, Michael. "Burroughs Adding Machine Company: Glimpses into the Past," http://www.dotpoint.com/xnumber/hancock7.htm (accessed September 2003)

Harrison, I. and S. Fossett. *The Book of Firsts: The Fascinating Stories Behind the World's Greatest Achievements, Discoveries, and Breakthroughs*. Pleasantville, NY: Reader's Digest, 2003.

I'll Buy That! 50 Small Wonders and Big Deals that Revolutionized the Lives of Consumers. Yonkers, NY: Consumers Union, 1986.

Kemmiya, Misa. "TED Case Studies, Shrimp and Turtle, Case Number 436," http:// www.american.edu/projects/mandala/ted/shrimp2.htm (accessedSeptember 2003).

Kwolek, Stephanie. Telephone Interview by Author. August 24, 2003.

Lillard, Margaret. "Re-Enactment of Wright Bros. Flight Fails," Associated Press, December 17, 2003.

McLaren, Carrie. "Porn Flakes: Kellogg, Graham and the Crusade for Moral Fiber," ibiblio: the public's library and digital archive, http://www. ibiblio.org/stayfree/10/graham.htm (accessed October 2003).

Mooney, Julie. *Ripley's Believe It or Not! Encyclopedia of the Bizarre, Amazing, Strange, Inexplicable, Weird and All True!* New York: Black Dog and Leventhal, 2002.

Papazian, Charlie. *The Complete Joy of Home Brewing*, 3d ed. New York: Harper Collins, 2003.

Rose, S. and N. Schlager. *How Things Are Made.* New York: Black Dog and Leventhal, 2003.

Schiavone, Louise "Australian inventor's gun fires 1 million rounds a minute" CNN, June 28, 1997, http://www.cnn.com/TECH/9706/28/super.gun/ (accessed October 2003)

Turkington. C. A. and S. J. Propst, MD. *The Unofficial Guide to Women's Health.* Boston, MA: IDG Books Wordwide, 2000.

Van Dorenstern, Phillip, ed. *The Portable Poe.* New York: Penguin Books, 1981.

WEB SITES

Access Cash
http://www.access-cash.com/us/programs/retail/anniversary_detail.html

American Plastics Council
http://www.americanplasticscouncil.org/index.html

Band-Aid Brand Adhesive Bandages
http://www.band-aid.com/index2.html

British Airways
http://www.britishairways.com/concorde/index.html

Capt'n Clint's Place

http://capt.clint.home.mindspring.com/index.html

Clonaid.com, Pioneers in Human Cloning
http://www.clonaid.com/news.php

Deka Research and Development Corporation
http://www.dekaresearch.com/index.html

Devil's Rope Museum
http://www.barbwiremuseum.com/

Dupont
http://www1.dupont.com/NASApp/dupontglobal/corp/index.jsp

Edward Lowe Foundation
http://edwardlowe.org/index.shtml

Enchanted Learning
http://www.zoomschool.com/Home.shtml

California Energy Commission, Energy Quest
http://www.energyquest.ca.gov/index.html

Carrier Corporation
http://www.global.carrier.com/

Drive-ins.com: The Definitive Resource for Drive-In Information
http://www.drive-ins.com

Etch-A-Sketch.com
http://www.etch-a-sketch.com/

Fender.com
http://www.fender.com

Goodyear
http://www.goodyear.com/

Forbes
http://www.forbes.com/

How Stuff Works
http://www.howstuffworks.com/

Goddard Space Flight Center, NASA
http://www.gsfc.nasa.gov/

The Great Idea Finder
http://www.ideafinder.com/

The History of Computing Project
http://www.thocp.net/index.htm

ibiblio: the public's library and digital archive
http://www.ibiblio.org

Independence Technology
http://www.independencenow.com/index.html

Invent Now
http://www.invent.org/index.asp

Inventors with Mary Bellis, at About, Inc.
http://inventors.about.com/

John Deere
http://www.deere.com/en_US/deerecom/usa_canada.html

Joseph Enterprises, Inc.
http://www.chia.com/

Kellogg's.com
http://www.kelloggs.com

Kentucky Educational Television
http://www.ket.org/

Lemelson-MIT Program
http://web.mit.edu/invent/

Levi Strauss & Co.
http://www.levistrauss.com/

The Library of Congress
http://www.loc.gov/

Lucidcafé
http://www.lucidcafe.com

Museum of Talking Boards
http://www.museumoftalkingboards.com/

National Fire Protection Association
http://www.nfpa.org/catalog/home/index.asp

The National Museum of American History, Smithsonian Institution, Fast
   Attacks and Boomers: Submarines in the Cold War
http://americanhistory.si.edu/subs/index.html

National Oceanic and Atmospheric Administration
http://www.noaa.gov/

National Park Service, Statue of Liberty
http://www.nps.gov/stli/

The Nobel e-Museum
http://www.nobel.se/index.html

Patently Absurd!
http://www.patent.freeserve.co.uk/index.html

Rothschild Peterson Patent Model Museum
http://www.patentmodel.org/default.aspx

Pez.com
http://www.pez.com/

Prozac.com
http://www.prozac.com/

Recovery Pharmaceuticals, Inc.
http://www.recoverypharma.com/index.html

Sealed Air Corporation
http://www.sealedair.com/index.htm

The Simon Lake Submarine Web Site 2003
http://www.simonlake.com

Thomson Delphion
http://www.delphion.com/

3M Worldwide
http://www.3m.com/

The Toxic Shock Syndrome Information Service
http://www.toxicshock.com/

The UK Patent Office
http://www.patent.gov.uk/

United States Patent and Trademark Office
http://www.uspto.gov/

University of New Mexico Health Sciences Center, Department of
  Anesthesiology and Critical Care Medicine
http://hsc.unm.edu/anesthesiology/

U.S. Food and Drug Administration
http://www.fda.gov/default.htm

Viagra.com
http://www.viagra.com/

Vinyl Record Day
http://www.vinylrecordday.com/

Wikipedia, the free encyclopedia
http://en.wikipedia.org/

Wilkes-Barre TWP. Police Department
http://www.wilkesbarretwppolice.org/

Zenith Electronics Corporation
http://www.zenith.com/

Zippo Manufacturing Co.
http://www.zippo.com/home.html

Photo Credits